Rural History in Europe 12

COST
Action A 35 PROGRESSORE

COST – the acronym for European COoperation in the field of Scientific and Technical Research – is the oldest and widest European intergovernmental network for cooperation in research. Established by the Ministerial Conference in November 1971, COST is presently used by the scientific communities of 35 European countries to cooperate in common research projects supported by national funds. The funds provided by COST – less than 1% of the total value of the projects – support the COST cooperation networks (COST Actions) through which, with EUR 30 million per year, more than 30,000 European scientists are involved in research having a total value which exceeds EUR 2 billion per year. This is the financial worth of the European added value which COST achieves.

A 'bottom up approach' (the initiative of launching a COST Action comes from the European scientists themselves), *à la carte participation* (only countries interested in the Action participate), 'equality of access' (participation is open also to the scientific communities of countries not belonging to the European Union) and 'flexible structure' (easy implementation and light management of the research initiatives) are the main characteristics of COST (Web: www.cost.esf.org).

As precursor of advanced multidisciplinary research COST has a very important role for the realisation of the European Research Area (ERA) anticipating and complementing the activities of the Framework Programmes, constituting a 'bridge' towards the scientific communities of emerging countries, increasing the mobility of researchers across Europe and fostering the establishment of 'Networks of Excellence' in many key scientific domains such as: Biomedicine and Molecular Biosciences; Food and Agriculture; Forests, their Products and Services; Materials, Physical and Nanosciences; Chemistry and Molecular Sciences and Technologies; Earth System Science and Environmental Management; Information and Communication Technologies; Transport and Urban Development; Individuals, Societies, Cultures and Health. It covers basic and more applied research and also addresses issues of pre normative nature or of societal importance.

Agricultural specialisation and rural patterns of development

Edited by
Annie ANTOINE

BREPOLS

This publication is supported by COST.

It is the result of the work launched in the Working Group 2, 'Management of Rural Land' of the COST Action A35 PROGRESSORE.

It came into existence thanks to the funding of both the ESF (COST), the Centre de Recherches Historiques (CNRS-EHESS) UMR8558 and the GDR (Groupe de Recherches of the CNRS) 2912, Histoire des Campagnes Européennes.

We are grateful to Anne Varet-Vitu (UMR 8558, EHESS-CNRS) who created the final lay-out of the book and to Judith Le Goff, who patiently edited all the English versions of the papers, to Paul Brassley who re-read the Introduction, and to Philip Hoffman who provided valuable help on several difficult technical questions.

Cover: *262 Kerscoul d'en haut* (2009), painting by Lucien Pouëdras (1937-). Oil on linen, 61 × 50 cm. Ecomusée de Saint-Dégan, commune de Brech en Morbihan, pays d'Auray.

© D/2016/0095/243

ISBN 978-2-503-53228-8

e-ISBN 978-2-503-55932-2

DOI 10.1484/M.RURHE-EB.5.106195

Printed in the E.U. on acid free paper.

CONTENTS

LIST OF CONTRIBUTORS

Annie ANTOINE	Université Rennes 2 France
Hrabrin BACHEV	Institute of Agricultural Economics, Sofia, Bulgaria
Magnus BOHMAN	Umeå universitet Sweden
Martine COCAUD	Université Rennes 2 France
Laurent HERMENT	Centre de Recherches Historiques, CNRS Paris France
Andreas KULHAWY	Stadtarchiv Peine Germany
Erich LANDSTEINER	Universität Wien Austria
Gérard LE BOUËDEC	Université de Bretagne Sud, Lorient France
Carin MARTIIN	Sveriges lantbruksuniversitet (Swedish University of Agricultural Sciences), Uppsala Sweden
John MARTIN	De Montfort University, Leicester United Kingdom
Pavel MATLAS	Jihočeská univerzita, Českých Budějovicích (University of South Bohemia in České Budějovice) Czech Republic
Benoît MUSSET	Université du Maine, Le Mans France
Pia NILSSON	National Historical Museums, The Archaeologists, Linköping Sweden
Matti PELTONEN	Helsingin yliopistossa (University of Helsinki) Finland
María Teresa PÉREZ PICAZO (†)	Universidad de Murcia Spain
Paul SERVAIS	Université catholique de Louvain Belgium

LIST OF FIGURES

LIST OF TABLES

The Series Rural History in Europe

Sometimes viewed as the producer of foodstuffs and raw materials, as a garden or a space for the production of energy; sometimes seen as a reservoir of manpower destined to be subverted to the needs of cities and industries or as a peaceful haven which welcomes weary city-dwellers; also regarded as a place of easily-incensed 'backward' societies or as an idyllic spot, close to nature where societies act as guardians of a certain conception of quality of life, the countryside has not ceased to be at the centre of economic, social, political, and now environmental concern of governments and public opinion. Pulled between contradictory demands, the countryside of today presents an image of a place and a society in clear transformation, which, in itself, is difficult to decipher.

In Europe, the difficulty is that the countryside, which forms a complex and evolving universe, is experiencing ruptures and also exhibits inertias. These complex transformations of the rural world can only be fully understood if they are viewed in a manner which transcends national boundaries and if the discrepancies, which can only be observed by adopting a broad view, in fact at the continental scale, are taken into account. These changes cannot be explained without reference to their past. The question is how to create such a dialogue between researchers which goes beyond national boundaries, crosses chronological barriers and breaks disciplinary boundaries.

The main objective of the 'Rural History in Europe' collection is to provide just such keys to unlock the changes experienced by present-day European rural societies in the light of their historical experience. It is to produce the historical knowledge which will allow us to conceptualise the future of European country-dwellers, as they face the kinds of problems which historians have grappled with in examining societies in the past: under-employment and multiple occupations, migrations and rural depopulation, distortion of competition by the marketplace or by the policies established by political authorities, competition with non-European producers, problems of resource allocation (land and water), distribution and redistribution of heritages and land-holding, the future of owner-occupied farms and the function of agriculture as employer, the tensions between private ownership and public access to land, the changes induced by settlement of urban migrants, environment and sustainable development, and so on.

In most European countries, rural society and long term change have been the subject of intensive research. Over the last half-century, a lively historiography has developed around questions concerning the factors of growth and systems of agricultural production, as well as the periodisation of growth and the impact of agriculture on industrial development. As a result, a body of knowledge has built up on techniques, forms of land-use, levels of production, the differing conditions and nature of the peasantry and the strategies of economic agents. More recently environmental concerns have made a dramatic impact on the consciousness of historians in the field, while previous findings have been questioned in a critical way and new perspectives have been drawn.

The field known as Rural History remains very much alive, despite the decline of the agricultural sector over the last several decades. Rural History remains fundamental in a Europe which has for so long been kept together by its Common Agricultural Policy and which now has to face the heavy impact of Brussels-based initiatives upon its rural regions, and an unprecedented revision of current agrarian systems and systems of production in the countries which have recently joined the EU. How can the changes taking place in present-day Europe be understood without taking into account a past which is still very present, and which determines both structures and behaviour?

The first volumes of this collection are the result of twelve workshops which have been held during three years (2006-2008) and which have mainly been supported by funds of the European Action COST A 35 and by other institutions like CORN, GDR CNRS Sociétés Rurales Européennes, Universities… The COST Action intends to extend the historical analysis of rural society over the last millenium in order to envisage the problems of the countryside in an extended timeframe and also to draw upon the commentary and expertise of specialists from other disciplines (sociology, economics, anthropology). This will enable the first real comparison of Europe from historians from all over the continent, from Scandinavia to the Mediterranean and from the western frontier to the eastern limits. Papers from these meetings will be published after a peer-review process, supported by the editorial board and, of course, the work of the authors and editors of the volumes. This collection will be constructed upon four main pillars: Landed property; The management of rural land; Peasant societies; The state, government, politics and peasants. Each of these will bring new perspectives and produce new tools to better understand the changes which are taking place today. In order to ask the relevant questions about the future of these peasantries and rural spaces in transformation, the books of this collection will deal with the *longue durée* and will present either research in progress or a synthesis on a regional or national scale. The cumulative effect of this approach will be to produce volumes, the geographical coverage of which will be the whole of Europe.

The volumes will also, of course, take into account the role of history as an explanatory factor for contemporary European Societies. If rural societies have been overthrown, if rural landscapes have been profoundly transformed and if the intervention of the State has considerably strengthened the regulation of production and trade, the contrast between a contemporary rural world in rapid transformation and a traditional rural world with frozen landscapes, petrified societies, immobile economies and lethargic political contexts will be an illusion. It is important, therefore, to detect, to measure and to interpret the range of recent changes by illuminating those that have taken place in past centuries in a European context.

1. Specialisation in rural history: towards a definition

Laurent HERMENT, Annie ANTOINE

I. Introduction

The idea of specialisation was late in arriving in the toolbox of rural historians. It was a loan from geographers and economists, and historians were slow to take possession of the possibilities that borrowing this concept offered them. For geographers as well as economists, specialisation is seen as a state, but also as a process: the abandonment of mixed farming (*i.e.* subsistence mixed farming) and the passage to some chosen form of production, because it was more profitable. This evolution can be observed at the level either of the individual farm or the region. Specialisation is considered as progress compared to the previous situation, and in return mixed farming is interpreted as linked to self-sufficiency, a consequence of the isolation of traditional farming areas. Specialisation is thus thought to be the child of developing means of transport, production techniques and the growing monetisation of rural economies, leading to the development of markets and trade networks. Isolated rural areas dedicated to subsistence mixed farming were followed by, at different periods depending on the country and region, systems based on production which often was unique, commercialised and highly profitable. In specialised writing on the twentieth century, specialisation often implies intensification, and growth both in the volume of production and in the efficiency of the system.

Although the agricultural specialisation phenomenon essentially concerns the contemporary period, during which it both accelerated and became widespread – which explains that the concept was especially employed by geographers and economists – it can nevertheless be studied over the long term. And it seems then to be a process which has its roots far earlier than the contemporary period. In France and England, mediaeval historians describe its first appearances as early as the mid-Middle Ages, in particularly favoured regions, those where large cereal farms were established by abbeys and secular seigneuries in the Île-de-France (Duby, 1962) or in the silt plains

Agricultural specialisation and rural patterns of development, ed. by Annie ANTOINE, Turnhout, 2016 (*Rural History in Europe*, 12), p. 13-45.

BREPOLS PUBLISHERS DOI 10.1484/M.RURHE-EB.5.112259

of Northern France (Derville, 1988). The first specialisation of European agriculture was indisputably in grain. Afterwards during the early modern period, agricultural specialisation spread to areas with no particular tendency (pedological and/or socio-economic) to produce grain, but with the ability to develop gardening on a large scale, such as Flanders or the South of England (Meuvret, 1953; Thirsk, 1992), or a system based on legumes and livestock, as in Norfolk (Beauroy, 1998). Finally, with the appearance of the railway, chemical fertilizers and mechanisation, the system became widespread and the rural world split in two: the modernised countryside whose farmers had defined specialisations guided by the marketplace, and outdated areas, mired in subsistence mixed farming.

The purpose of this book[1] is to try to reinterpret the antecedents to contemporary agricultural modernisation in light of the heuristic possibilities offered by the notion of specialisation. In farming defined as 'traditional' (as opposed to modernised agriculture), without more precision in view of the chronological gaps which split the area considered in this publication, self-sufficiency and home consumption characterise practically all farms and all systems of production to varying degrees. In most of the agricultural systems mentioned in this book, a part of the production is eaten/used by the farm itself. This is the particularity of traditional agricultures, as Richard Cantillon described in 1755 when he exposed the theory of the 'three rents' of a farm. 'It is a common idea... that a farmer should make three rents' (Cantillon, 1755): the one he pays to his landlord, the one he uses to support himself and his animals, and the one 'which must remain for him to make a profit for his farm.' The first corresponds to the rent for the land, the second represents home consumption (feeding people and animals), and the third to the money and the share of production reinvested in the operations of the farm. And if it is possible in spite of this to imagine and even to point to farmers who were in no way involved in the marketplace, but only produced for food and ate everything they produced, this was evidently never the general rule. The model described by Cantillon in the early eighteenth century which led a large percentage of farmers to intervene in the marketplace was evidently the most plausible because, at some time between the Middle Ages and the contemporary period, it was necessary for the peasant to have cash available at least to pay his taxes, probably his manorial rights, and often the rents for his land. The peasant thus had to sell, whatever the size of his farm. It is in this part of his production is that it is worth looking for forms of specialisation, on the fringes of the part consumed by the

[1] In June 2007, a workshop was organised in Rennes (Agricultural Specialisation and Rural Patterns of Development) as part of COST A 35. It was devoted to the study of agricultural specialisation and rural patterns of development from the Middle Ages to today. It was based on the assumption that agricultural systems have not necessarily been averse to any form of agricultural specialisation, but that it can only occur adjacent to the main goal of agriculture, which has traditionally been to feed the farm household. This volume was written using texts presented at that workshop.

household. Let us note that the two networks were not hermetic but intertwined, and they supported each other as it were, since some of the production (such as fodder, straw and fertilizer) was integrated into the commercialised part.

The specialisation of traditional agriculture was multifaceted, for it developed from the diversification of activities. It could focus on one type of production, but was most often made up of several linked activities. Specialisation was not then necessarily synonymous with monoculture, even if there were sometimes examples of it, in the case of grapes or certain dye-plants such as madder, on traditional farms. It is paradoxically in this kind of production, often described in a way which is indistinguishable from mixed farming, that specialisation can be observed. Indeed, diversity of production was inherent in farming as long as technical development remained weak (as, for example, in ploughing and pest and disease control). The traditional farmer, with little mechanical or chemical assistance, had to diversify his production. His survival was at stake in an era in which a harvest could be destroyed by weather, pests or diseases, so he was better off with a range of products. Thus, in traditional systems, mixed farming was the general rule. But mixed farming can mean very different things in different contexts of farm size, land distribution, landlord/tenant relationships, market organisation, and outlets for agricultural products. It can be entirely or partially subsistence mixed farming, but can also imply a diversification of production which leads to farm specialisation. It will then be through this diversification that specialisation will first be observed.

To do this, it is worth defining the meaning of diversification as precisely as possible. We will therefore propose a definition for specialisation. This work, which was in part inspired by the theories of business economics (or organisational economics), constitutes a necessary heuristic detour in order to better understand the typology of agrarian systems as seen through the prism of the specialisation/diversification dichotomy. This detour in no way signifies that the farms of the early modern period and the nineteenth century were comparable to the giant firms that organisational economics (management) preferentially brings to mind. It is an effort to put into perspective the conceptual pair of diversification/specialisation that rural historians struggle to define.

We will first examine the contributions available in this volume in the light of some classic works on rural history with the help of the theory of Von Thünen, who was the first to have tried to examine the geographical distribution of agricultural specialisations. Secondly, we will sketch a conceptual outline of the notion of specialisation in agriculture from the early modern period and the nineteenth century through the *capability* concept put forward by G. B. Richardson (Richardson,

1972). Thirdly, we will envisage the same problem from the angle of Transaction Costs Economics (henceforth, TCE), by revisiting the presumed founders of this theory (Williamson, 1985). With the help of these concepts, we will show that it is appropriate to envisage two types of problems: the vertical integration of the process of agricultural production and the horizontal diversification of production. This long heuristic detour will lead us in conclusion to propose an outline which we hope will make it possible to understand methods of specialisation on large and small farms.

II. Theoretical approaches to the notion of agricultural specialisation. The effects of spatialisation

Igor Ansoff, drawing inspiration from the Queen of Hearts in *Alice's Adventures in Wonderland* indicates that 'Just to retain its relative position, a business firm must go through continuous growth and change. To improve its position, it must grow and change at least *twice as fast as that*.' Ansoff adds 'There are four basic growth alternatives open to a business. It can grow through increased market penetration, through market development, through product development, or through diversification...diversification decisions...much more than other alternatives,... require a break with past patterns and traditions of company and an entry onto new and uncertain paths.' (Ansoff, 1957: p. 113).

In a moment of introspection, the rural historian might wonder how the farms of the early modern period and the nineteenth century managed to diversify their production without subjecting themselves to the law of the Queen of Hearts. Indeed, until the end of the nineteenth century, agrarian systems were characterized by a very great diversification. It even seems that, unlike industry, specialisation was what raised the most problems. But it should be pointed out that the concepts of specialisation and diversification have not always been clearly defined in the literature dedicated to this topic[2]. This semantic wavering is not unique to rural history. In management science as well these terms can lead to confusion. Ramanujam and Varadarajan noted in 1989 that a 'review of the literature reveals that there is a great deal of variation in the way diversification is conceptualized, defined, and measured.' (Ramanujam, Varadarajan, 1989: p. 524).

To keep to rural history, it must be acknowledged that the theoretical approaches of these two notions of specialisation and diversification are rare, even non-existent. The analysis of two landmark works illustrates these uncertainties. That of Joan

---

2 It is symptomatic that the term specialisation does not appear in the indices of works on rural or agrarian history, including general works like that of G. Federico (Federico, 2005).

Thirsk, published in 1997, is revealing. While demonstrating that agrarian systems were susceptible to adaptation, her arguments shed little light upon this vocabulary question. According to this author, all agricultural activities not centred on grains or livestock are 'alternative agricultures' (Thirsk, 1997). Should it be concluded that grains and/or livestock constituted the first forms of specialisation, or that the agrarian systems based essentially but not necessarily exclusively upon one or the other of these productions were diversified systems[3]? Alternatively, is it necessary to conclude that all farms which did not essentially practise one or the other of these two activities were specialised farms?

One of the chapters of the book by Jan de Vries on transformations in the agriculture of the northern Netherlands, published in 1974, explicitly echoes the problem raised here. After having noted that it was difficult to evaluate the changes in the agriculture of the seventeenth- and eighteenth-century Netherlands, Jan de Vries notes that:

> One aspect of the transformation stands out clearly: the growth of scale. Rural households reorganized their activities; they shed from their work schedules a wide variety of tasks necessary to sustain the household in a regime of relative self-sufficiency and concentrated their efforts on the remaining tasks, the more strictly agricultural tasks. In a word, they specialized. An indicator of specialization is the growth of the scale of production, and in areas predominantly engaged in livestock husbandry, the growth of herd size shows this development most clearly. (de Vries, 1974: 136-37)

According to Jan de Vries specialisation implies the growth of farms' activity level and the areas controlled by the same farmer[4].

Evidently, the notion of specialisation and the specific nature of the specialised farm have not been clearly defined by historiography. For Joan Thirsk, the notion of specialisation, if it has any meaning at all, refers to types of products. For Jan de Vries, the notion refers to the economic model of farms. It would be easy then, based on the second approach, to associate specialisation with the large farm and diversification with the model of self-sufficiency which characterised the small, even the very small, farm[5]. Ultimately, this absence of clear theoretical foundations is even more surprising because the historiography of the rural Europe usually describes the endlessness of specialised agrarian systems: there are numerous regional studies

---

[3] For a nuanced, very useful critique of Joan Thirsk's book, see Poussou, 1999. For France see Béaur, 2009.

[4] In fact, specialisation does not necessarily mean growth in area. For a particularly characteristic example, that of California at the turn of the twentieth century, see Olmstead, Rhode, 1993.

[5] For Michel Gervais, Claude Servolin and Jean Weil, for example, small and medium-sized French farms would be characterised, until well into the twentieth century, by a large dispersion of production, although the large farms, those of the greater Parisian Basin in particular, would concentrate all their efforts on a few lucrative enterprises (Gervais, Servolin, Weil, 1965).

which insist on the specificities of local production, the originality of farming cycles and of the networks of commercialisation designed to sell the products.

In turning to spatial analyses of the zones of agricultural specialisation, the question does not necessarily appear to be posed any more clearly. Yet Von Thünen's theory seems to offer a practical enough means for interpreting the geography of agricultural specialisations and the conditions of the commercialisation of agricultural products[6]. Production was organised in concentric circles around urban hubs. In this pattern, illustrated numerous times, market gardening and fresh dairy products settled in towns' immediate peripheries. Further out, agricultural land was devoted to cereals. Finally, livestock-rearing regions were sometimes situated several hundreds of kilometres from where the animals were consumed[7].

However, relationships between agricultural activities and their commercial outlets rarely followed Von Thünen's model, for different reasons which do not all have the same impact. First of all, in this model, all locations equidistant from the city-state were supposed to develop the same types of production because of the uniformity of the soil and the homogeneity of the space. In fact, as shown by Jean Meuvret for France, geography imposed its constraints. The smallest difference in level was an obstacle to circulation (Meuvret, 1988). In contrast, the existence of a canal network, a navigable river, or proximity to the sea facilitated the flow of products. In this volume, Erich Landsteiner demonstrates that the problem did not only concern wheat but also wine. Secondly, the evolution of techniques must be taken into account. The situation of England after 1870 illustrates the utility but also the limits of Von Thünen's model[8]. As John Martin recalls in his contribution, in the interwar period, the whole world was called upon to supply English cities[9]. The drop in the costs of freight brought about by the appearance of refrigerated transport disrupted supply conditions and provoked a redefinition of regional agricultural specialisations around the world[10]. Thirdly, to explain the geography of trade, the context must be taken into

---

6 As the complete edition of the Von Thünen's work is not available in French, we based ourselves on Jean-Marie Huriot's book which provides large extracts and a stimulating commentary (Huriot, 1994).

7 On market gardening around Paris in the sixteenth and seventeenth centuries, see Gurvil, 2010. On the commerce of oat headed for urban cavalries around Paris in the eighteenth century, see Moriceau, 1994. On the raising of cattle destined to supply Paris with meat in the early modern period and the nineteenth century, see Delhoume, 2009 and Garnier, 1999. For the commercialisation of agricultural products in England 1750-1850, see Perren, 1989. See also recent studies by Mikael Kopsidis, Nicolaus Wolf and Merijn Knibbe which perfectly illustrate the influence of this model on Germany and the Netherlands. Kopsidis, Wolf, 2010; Knibbe, 2011. See also Kopsidis, 2009.

8 Long before England, cereal supplies to the Netherlands depended on foreign sources (de Vries, 1974: 169-173)

9 On the question of supplying English cities in the nineteenth century, see Scola, 1992.

10 For the specific analysis of these questions and the limits of Von Thünen's contribution, see Huriot, 1994: 275 and after. On the organisation of the marketplace and a critique of Von Thünen's model which

account. To take an example borrowed from France, during the period of frumentary crisis, Paris's supply zone, which normally covered nearly 40,000 km^2, expanded to the European scale (Bourguinat, 2002: p. 113-148).

It is worth emphasising that distortions of the agrarian space linked to context, physical geography or the evolution of techniques would not, by themselves, be enough to call into question the validity of Von Thünen's model. If the geographic distribution of specialisations do not obey Von Thünen's model, it is because in last analysis, this distribution was controlled by the establishment of complex commercial networks, which led to the organisation of supply channels. Kevin O'Rourke and Ingrid Henriksen forcefully remind us of that in their study of the emergence of the Danish butter sector (Henriksen and O'Rourke, 2005). The complexity of the relationships between agricultural specialisation and outlets also refer in many cases to the organisation of industries which used agricultural raw materials. From this point of view, several models are possible which do not willingly yield to Von Thünen's pattern. We will content ourselves with two scenarios here: first, sugar beet, as indicated in the contribution of Andreas Kulhawy, whose expansion was largely determined by the establishment of an industry designed to process agricultural raw materials. Next, flax and hemp, whose success depended on the establishment of a proto-industrial system and an efficient commercialisation network (Ronsijn, 2013)[11].

The relationship between agricultural specialisation and the establishment of commercial networks ultimately concerned all types of production. In this volume, the contributions of Carin Martiin, Matti Peltonen and Martine Cocaud highlight the dependence of butter producers on commercial channels. This dependence could take several forms and could be more or less restrictive. At the end of the nineteenth century, Finnish producers had the choice between a quality product destined to be exported, and a product destined to supply local markets. The modes of developing the two types of products did not obey the same principles: the prices of products destined for export were determined in Copenhagen, while the much lower price of the products destined for the local markets was decided upon locally. The two commercialisation networks did not operate on the same scale, and did not obey the same mode of regulation. According to Martine Cocaud, in Brittany in the nineteenth century, the introduction of small farmers into the butter market would have depended much more on the choices made by merchants than by the peasants' decisions. The

emphasizes the dominant relationships of the merchant city over its outskirts, see Braudel, 1979: 27-28, and Braudel, 1982: 38-39 for the English edition.

[11] Concerning this last example it is utterly impossible to cite the multitude of works that mention it. Pia Nilsson demonstrates, through the example of the production of hops in seventeenth-century Sweden, that the geographic location of farms in relation to urban centres can not in itself explain the processes of specialisation.

mediocre quality of Breton butters, which were exported as far as the New World in the eighteenth century, ultimately would not have constituted the cause for the decline in butter specialisation. To explain this decline, it would be necessary, according to the author, to blame the attitude of the intermediaries.

It can be seen through these examples that specialisation choices did not obey pure geographical determinism, but closely depended on the existence and the policy of commercial networks over which peasants often had little control. The Huerta of Murcia example in the early modern era demonstrates that the peasants could be doubly imprisoned. On one hand, the choice of specialisation was the responsibility of the landlords. On the other hand, they depended entirely on intermediaries who marketed their products.

The phenomena of specialisation cannot be reduced to Von Thünen's model. It is also worth analysing them in terms of diversity and complementarity, as Wouter Ronsijn points out (Ronsijn, 2013). Diversity and complementarity of markets first: goods, services, work, land and credit. Diversity and complementarity of farmers' situations second: the position of large famers was not the same as that of small farmers, the latter was not the same as that of the rural workers who had only their labour to sell, but for whom access to markets was absolutely vital. Ultimately, all classes of the peasantry wished to have access to markets, not necessarily as hyperactive merchants thirsty for profits, even if this way of thinking was not foreign to them, but because markets were the way for their activities to spread[12]. Thus it is worth questioning the diversity of institutional forms that this relationship took. But firstly the working of the diversification/specialisation pair must be defined more precisely.

III. Diversification and specialisation. Notions of 'pivot-production' and the production system

It is firstly worth recalling the necessary diversity of production in traditional agrarian systems, illustrated numerous times by the contributions in this volume (see particularly Martine Cocaud, Pia Nilsson, or Pavel Matlas) if only because peasants most often had to assure their own survival and always worried about not having enough (Meuvret, 1988: 146). But other factors also explain this diversity. The first refers to a trivial economic calculation: not putting all one's eggs in the same basket. If a harvest was bad, or if a product hit a slump, farmers imperatively had to have alternative products. Additionally, ecology and pedology imposed

[12] For pre-industrial France, see Margairaz, 1988. For seventeenth-century Germany, see Warde, 2006. For the Netherlands in the early modern era, see de Vries, 1974: 155-164.

often insurmountable constraints, given the state of technology. Among others, the contributions of Magnus Bohman, Pavel Matlas, Maria Teresa Pérez Picazo, Pia Nilsson and Martine Cocaud bear witness to these priorities. Hence the emphasis which for the last thirty years has been put on the ecological constraints governing traditional agrarian systems[13]. Does this then mean denying the possibility of absolute product-specialisation within traditional agriculture? The answer is clearly negative. However, as the contributions of Benoît Musset, Maria Teresa Pérez Picazo or Erich Landsteiner aim to demonstrate, this economic model more particularly concerned small or even micro farms[14]. It is not an adequate model to understand all the production diversity in traditional agrarian systems. It is especially unsuitable for understanding the economy of the largest farms.

Next it is worth differentiating the diversity of agricultural production from the diversification process of agricultural production. The diversity of production is not in itself a clue to a lack of specialisation. Speaking of the diversity of productions is neutral from the point of view of farm performances. It has been seen that mixed farming, which by definition implied a rather large diversity of production for a farm or a region, was not generally seen as economically reliable if this diversity corresponded to output mostly consumed by the household. But the diversity of productions at any given moment and in any given place could also be due to a desired process of diversification, and in that case is synonymous with specialisation based on a pre-existing model. In brief, it could be said that diversity of production is a state which in theory was neither negative nor positive from the point of view of the efficiency of a farm, but that diversification was a process that led to specialisation, a phenomenon that is generally interpreted in terms of economic progress of a farm or region. It is therefore better to think about the diversification/specialisation pair rather than each term separately.

To develop this thought, we will start with three distinct levels of analysis of this pair. The first refers to products, the second to the relationships that the farm and/or the farmer had with markets, and the third to the production process.

III.1. A necessary diversity of activities

The first level of analysis simply refers to the diversity of farm products. This first approach is doubtless the least fertile. At the end of the eighteenth century, a

[13] Among recent work pertaining to this question, see Clar, Pinilla, 2011. In France this is a relatively old question. Jacques Mulliez's very important article testifies in its own way to these preoccupations among rural historians (Mulliez, 1979).

[14] For a particularly enlightening example, see Niels Grüne's work on southwestern Germany (Grüne, 2007).

large farm in Norfolk produced wheat, barley, oats, turnips, etc. It also maintained a large amount of livestock (cattle, horses, and sheep) meant to be fattened (cattle and sheep), to work on the farm (horses, draught animals), or finally for wool (sheep). Was this farm specialised, and if so, how to measure to what extent? In comparing this farm to a large farm of the Parisian Basin from the same period, it seems that the latter was much more specialised than the former because its prosperity mainly if not exclusively rested on three products: wheat, oats, and to a lesser extent, sheep. There were no turnips or cattle, only a few cows. The farms of the Parisian Basin only very gradually made room for legumes and meadows (Moriceau and Postel-Vinay, 1992; Moriceau, 1994; Kim, 1999; Herment, 2009 and 2012). All in all, in terms of diversity of production, the farms of Norfolk were much closer to the farms of the Limousin, which Turgot still decried as backward (Delhoume, 2009). How is it possible to measure the relative degree of specialisation of farms whose economic models were so different? Evidently it is worth abandoning the definition of specialisation in terms of products. Only in focusing the analysis on the relationships of the farm to the markets on one hand, and in taking into account its entire production process on the other, will it be possible to understand its level of specialisation.

These examples, and those given in this volume, clearly show that the degree of the product diversity did not in itself constitute a positive indication for a farm or region. The same degree of diversity can point to a very archaic mixed farming system since it is fundamentally linked to home consumption (producing everything to be eaten and perhaps selling off the surplus, maybe responding to market demand), signal farms which had multiplied their productions to circumvent the risks inherent to the production process (in a technical context which made it impossible to avoid plant and animal diseases, parasitic attacks, etc.) or demonstrate the workings of the marketplace (in an economic context where the producer did not have a dominant role and was rarely the one who set prices). In this latter case, it triggered a modernisation process which appears more clearly in using the second definition of specialisation.

III.2. Integration in markets[15]

Specialisation can be seen as a consequence of the integration of farms in the markets. This second level of definition seems more fertile than the first. From this point of view, as Carin Martiin and Pia Nilsson remind us, specialisation assumes the development of a product whose volume exceeded the needs of the household and which therefore became marketable. This approach potentially contains several advantages.

[15] For a broader understanding of the relationship between agricultural growth and markets, we recommend reading the first chapter of the second volume of the collection *Rural History in Europe* (Pinilla, 2009: 11-36).

Firstly, it offers a practical way to measure the degree of specialisation of farms. It would suffice to link the value of each product to the total value of the farm's production and to isolate the marketed products in order to measure their relative importance. Yet this simple, seductive method makes little sense. In the type of agriculture defined by convenience as traditional, a very large percentage of products were consumed at home. It is not a question of identifying traditional farms and subsistence agriculture, it is a question of admitting that a division between marketable production and non-marketable production did not necessarily make sense, since whole sections of the production were reinjected into the productive process, even if the farm's products were largely commercialised. Straw, hay, manure, wood, and the grain consumed by people and animals can be cited in no particular order[16]. In this volume, Matti Peltonen's contribution shows the importance of products consumed at home in Finland from the late nineteenth and early twentieth centuries, including the most marketable products. Hrabrin Bachev also brings up the question in a contribution concerning Bulgaria from 1990-2005, or again Martine Cocaud for the bocage regions of western France in the nineteenth century.

Concerning this question of the reintroduction of a part of the production in farms' productive process, it is worth highlighting the fact that before artificial fertilizers were used on a massive scale, farms were subject to a practically insurmountable ecological constraint. It was imperative to give back to the soil the fertilizing elements that the crops and livestock had removed[17]. And to do that, farmers could only count on their own efforts. Reinvesting straw, manure and ashes among other things was essential. However two types of farms managed to benefit from large amounts of fertilizer: on one hand, farmers located in zones which were densely populated or close to large urban centres, managed to loosen this constraint by buying urban sludge[18], on the other hand, as Gérard Le Bouedec recalls in this volume, peasants in coastal regions could use massive amounts of kelp. However, outside of these two specific cases, it was not until the seventeenth or even the eighteenth century that farmers succeeded in loosening this constraint. The introduction of meadows and the

[16] French textbooks on agricultural accounting from the nineteenth century report on the problems brought up by these phenomena in terms of the evaluation of inputs (LEMARCHAND, 1993: particularly from p. 354 onward; HERMENT, 2012).

[17] It is worth emphasising the significance of this massive expansion in fertilizerproduction and consumption after 1945-1950. In France, for example, the question of home consumption of products, still current in the early 1950s (CAMBIAIRE, 1952), became an anecdote, even folklore, firstly with the total dissociation of breeding and crop production in poultry farming, then in other domains (GERVAIS, SERVOLIN, WEIL, 1965: 105). In Denmark, this occurred as early as the end of the nineteenth century (HENRIKSEN, O'ROURKE, 2005). In certain zones of intensive breeding, it then provoked an overflow of fertilizers that agriculture was incapable of recycling. See the example of pig farming in Aragon (CLAR, PINILLA, 2011).

[18] For this topic see the work of Liam BRUNT, 2007.

growth in area dedicated to legumes constituted the first breach in the old system. The problem of nitrogen enrichment was temporarily and partially solved[19]. For phosphate, the obstacle was not removed until the appearance of the first artificial fertilizers (guano and bone char i.e. carbonized bone powder from sugar refineries)[20].

Secondly, the analysing specialisation in terms of market relationships explains the choices of specialisation. According to the principles of economics, economic agents are better off developing the most profitable products. With this in mind, choices of specialisation would have been guided by a calculation of maximising net revenue, which depended on the relative prices of different products and inputs. (Among the numerous recent studies based on such an issue see Knibbe, 2011; Fonseca, Reis, 2011. See also in this volume Magnus Bohman's contribution.) But ultimately such a view conforms rather poorly to the working of farms and traditional agrarian systems. As Carin Martiin writes in this volume, 'early stages of diary specialization in Sweden may not be interpreted as a mirror of grain prices.' On one hand, productive choices were largely controlled by ecological, institutional, and legal imperatives which it is difficult to reduce to the imperative of maximising net revenue. The assumption of autonomy of agrarian systems, and even more so individual farms, quickly finds its limits from this point of view. On the other hand, the different rhythms to which farms were subjected, reduced the possibilities of instantaneous adaptation of production to changes in the price structure[21]. Of course, all organisations were confronted with the same problem, but farms were particularly exposed, since the processes of production were controlled by biological rhythms which were irreversible in the short or medium term, even in the long term in certain cases. It was not only a question of a problem linked to the irreversibility of strategic investments, which is one faced by all types of businesses. It was a question of biological and ecological processes which concerned the products themselves (Servolin, 1972; Moser, Varley, 2013). Even in the cultivation of annual plants, like grains, the farmer had to take into account the constraints linked to successive crops, the organisation of crop rotation. The breach in crop rotation, which occurred in French Hainaut at the end of the seventeenth century in order to increase wheat production, could only be conceived on a small scale, except to

[19] The introduction of meadows is in fact much older. The question of knowing when, where and how their extension radically modified the conditions of enhancing the soil is not the objective of this volume. It will simply be noted here that, depending on the zone, their effects were sometimes signaled as early as the seventeenth century (de Vries, 1974: 142) or much later (Moriceau, Postel-Vinay, 1992; Herment, 2012). In other regions, such as south-central Portugal, the climate did not permit their introduction (Fonseca, Reis, 2011).

[20] For France, see Bourrigaud, 1994: 143 onward.

[21] This does not signify that the choices of specialisation were independent from the price structure, but that in the short term, the flexibility for adaptation of agrarian systems was relatively limited. This was obviously not the case in the medium or the long term, as the contributions to this volume indicate.

return to the soil the fertilizing elements that the intensification of crops had removed (Delleaux, 2012).

The approach in terms of the relationship to markets is nevertheless useful in that it makes it possible to understand an essential element in the application of a diversification/specialisation process: the choices made could not have existed without outlets. Lastly, the existence of outlets was in some ways a prerequisite, but only a prerequisite, to the application of a diversification/specialisation process.

III.3. 'Capabilities', pivotal products, and product systems

The third approach, pointing to the balance of productive processes, is undoubtedly the most fertile as it allows on one hand an examination of the specificities of the diversification/specialisation processes in traditional agrarian systems both synchronically and diachronically, and on the other an understanding of the limits of these specialisation processes. In order to better grasp the interest of this third approach, from organisational economics we will borrow the concept of 'capability' that G. B. Richardson put at the centre of his analysis of the firm.

Richardson defines an organisation's 'capabilities' as 'appropriate knowledge, experience and skills,' and notes that 'what concerns us is the fact that organisations will tend to specialise in activities for which their capabilities offer some comparative advantages.'(Richardson, 1972: 888). Obviously it is not a question of indiscriminately pasting this concept onto farms, but re-examining the notion of specialisation in agriculture based on a new point of view. Additionally this notion is particularly useful as it puts alternative modes of 'co-ordination' of firms and their relationship to markets back in the centre of the analysis[22]. It will also make it possible to re-examine the notion of specialisation while taking into account a phenomenon which previous developments had foreseen: vertical integration. Indeed in traditional agrarian systems, diversification of production did not principally manifest itself in horizontal diversification, but seemed to have obeyed a phenomenon of vertical integration.

In examining the diverse types of specialisation mentioned by Paul Servais in his contribution, it seems that at the end of the nineteenth century, at the end of a century of evolution, the 'capability' of small farmers in the Pays de Herve was to produce butter and cheese, and that of the large farmers in Hesbaye was to produce beet for the sugar industry. This issue, centred around 'capability', thus makes it possible to grasp the reasoning behind the processes used by farmers, their evolution and the constraints they faced.

[22] Richardson indicates that 'co-ordination can be effected in three ways: by *direction*, by *co-operation* or through *market transactions.*' (Richardson,1972: 890).

Let us again take the example of the large farms in Hesbaye and Brunswick which Paul Servais and Andreas Kulhawy respectively discuss. From the beginning to the end of the nineteenth century, a system of production centred on wheat shifted to a system of production based on sugar beet. This shift in the centre of gravity of large farms, common to many areas because it was found in Picardy (Postel-Vinay, 1974), cannot be limited to a reduction of the area dedicated to grains for making bread (wheat and rye) and a growth in area dedicated to beet. It led to total reconstruction of the economy of farms which, in France and Belgium, provoked the decline of sheep-raising as well as the appearance of working cattle and fertilizer, and in Brunswick, the transformation of cattle-raising through the introduction of permanent housing. In certain regions, it was accompanied by a growth in the area dedicated to oats in order to respond to the internal demand of farms in terms of draught force and the demands of urban horses. Therefore, far from being reduced to monoculture, or specialisation in terms of product, this evolution in farm capabilities entailed the emergence of a new agro-productive system, the appearance of new products and finally the growth of certain traditional products, which were just as much similar as complementarity activities, to use Richardson's terms[23].

Ultimately, a farm's 'capability' was in some ways the pivotal crop or a group of pivotal crops, around which its entire production process was organised. This analysis also makes it possible to explain one of the dominant traits of large traditional farms which, contrary to what one might think, were not characterised by a very high degree of specialisation in terms of product. On the contrary, it was the diversity of production that best characterised them. Proof of that is the multiplicity of products of the large farms in Norfolk which, far from restricting themselves to cereal crops, diversified production thanks to a complex system of crop rotation[24]. This analysis also goes for the farms of merchant-peasants in the northern Upper Rhine Valley which, from the end of the seventeenth to the middle of the nineteenth century, developed a complex mixed farming system resolutely directed at the market (Konersmann, 2011). This was true again for the farms of Bas-Maine in the eighteenth century (Antoine, 1994; 1999) or those of Brittany in the nineteenth century, as indicated by Martine Cocaud's contribution. It doubtless explains in part the absence of specialisation on large farms in southern Bohemia mentioned in the contribution of Pavel Matlas. The question was then not one of the diversity of products but about a pivotal crop around which farms' economic model was centred[25].

---

23 'I argue that organisations tend to specialise in activities which, in our special sense of the term, are similar. But the organisation of industry has also to adapt itself to the fact that activities may be *complementary*.' (Richardson, 1972: 889).

24 On the evolution of productions in East Anglia in the seventeenth and eighteenth centuries, see Holderness, 1984.

25 In her contribution Maria Teresa Pérez Picazo employs 'vector crop'. The term pivotal crop

This notion of pivotal crop(s) deserves to be analysed in detail because it makes it possible to greatly advance in defining the concept of specialisation. It has the advantage of being able to be equally applied, although in a diversified manner, to large and small farms. The pivotal crop could have been a single product (such as wheat, sugar beet or maize on large farms, as well as grapes, woad or hops for example on smaller farms) and there we are dealing with specialisation in the form of monoculture. But the notion of a pivotal crop should often also be understood as a shortcut designating a group of linked products, either that the former served to produce the latter, or that the revenue made from the various products combined to insure the farmer's economic security. This can be demonstrated by several examples. On tenant farms widespread throughout Mediterranean Europe since the Middle Ages and still at the beginning of the twentieth century, production was greatly diversified: cereal, legumes, grapes, olives, blackberry bushes for raising silkworms, hemp, flax, etc (Biagioli, 1989; Congost, Jover, Biagioli, 2003; Saguer, 2011). One part was for home consumption, but another part – all of the landlord's share and some of the tenant's – was put on the market. Here was a mixed productive system of which one part of the production was specialised. The case of farmers in northern Brittany can also be mentioned. In the middle of the nineteenth century they still produced rye and buckwheat to eat but sold the wood from pruned hedges, the milk from cows after calving, cider made with apples harvested from planted fields, sometimes even wheat if they had marine fertilizer or naturally favourable soil conditions (Antoine, 2015). Their farms were small or mid-sized and they were characterised by a large diversity in their products, however it is possible to speak of their diversification – the adaptation to the regional market orientated the choice of products sold – and thus specialisation. It looks like mixed farming, but a type of mixed farming in which only one part is subsistence (cereal to feed the farmer) and another made up of diversified production, the whole forming a specialised productive system. And if the diversification and the organisation of the productive systems of several types of small and mid-sized farms from the South to the North of Europe are scrutinised in the same way, it is clear that specialisation can frequently be brought to light. Ultimately, it appears that a majority of farms with very diversified production were very far from a subsistence model of mixed farming, but demonstrated traits of specialised farms.

To identify specialisation, it is worth examining and analysing the farming system – in the sense that, in *Agrarian History*, the 'wheat-sheep system' would characterise

is sometimes used to designate the crop which was at the head of rotation. For the purposes of this introduction, this term is meant differently although it often happens that the two are confused. In certain cases the definition of the pivotal crop or activity can be questioned. For Norfolk from 1650-1750, on can hesitate to choose one product or another, but if a choice has to be made, it would have been barley or wheat (Holderness, 1984). In Kent, depending on the region and the farm, it would have been wheat, livestock, barley or hops.

the openfields of the Midlands – to find out if a farm was specialised. These farming systems could be of different types. It could be a question of products which were linked because their association allowed for more productive soil, like in the la coltura promiscua case for example. It could be commercialised products integrated in simple markets (products with a direct, most often local, outlet) or in complex markets (links between local and international trade). But these products could also have been absorbed by markets made up of several interconnected segments which are not always easy to distinguish, especially for the early modern period when the origin of the product is rarely mentioned in the documents. Wheat from Dantzig arrived in France in the event of a large wheat shortage, and when it passed by the Sound, it was taxed as Dantzig wheat, but what was its initial origin? Probably a farm in Central Europe, but where exactly? (Pourchasse, 2006). This was the case of wines from all of Southern Europe, cattle from Maine, Normandy or the Massif Central, as found after several exchanges on the Parisian market, mules from Poitou which crossed the Pyrenees after having passed through several hands. The examples could be multiplied to infinity, and all these products were from farms which could be considered specialised. Finally these productive systems indicate that specialised farms, whatever the size and the economic performance, could be based on agricultural products feeding the market for raw materials for crafts (such as wool, flax and hemp) or what are called industrial cultures in the nineteenth century.

For small farmers, this choice of product-specialisation made it possible to generate a monetary profit by using the family workforce intensively (in this volume, see the two contributions of Erich Landsteiner and Benoît Musset). Finally, given the tiny size of many micro-farms, this choice did not prevent the use of family labour off of the farm. This is the mechanism that Benoît Musset describes with precision in his contribution. Here it is worth reasoning more in terms of pluriactivity than mixed farming. Gérard Le Bouëdec also provides a magnificent example of this phenomenon through the analysis of the case of wine-growing mariners in eighteenth-century Provence. Here, it can be surmised that wine was only an accessory, and that the maritime business constituted the pivotal activity. In Brittany, on the contrary, it seems that the sea was only an accessory to the land. At the two far ends of Europe, Maria Tereza Pérez Picazo and Matti Peltonen offer a particularly clear vision of the importance of pluriactivity in Huerta de Murcia in the nineteenth century or in Finland at the beginning of the twentieth century[26].

The concept of pluriactivity must also be analysed in order to show its differences and its tangency with the diversification and specialisation of farms. Firstly it is

---

26 For other examples of pluriactivity concerning Paris's sphere of influence in the early modern period and the nineteenth century, see Gurvil, 2011 and Herment, 2012.

important to observe that unlike the latter two ideas, the notion of pluriactivity has been conceptualised relatively well. It rests upon the idea that for small farmers, craft activity or belonging to a proto-industrial network supplemented family revenue at the same time that it used surplus labour. It has been greatly studied by historians of craftsmanship and proto-industrialisation (Deyon and Mendels, 1981: 11-19) without much attention to the rural aspect of these peasant craftsmen who spun wool (England, Scotland) wove cloth (all over early modern Europe), made pottery (Southern England), hosiery and lace (Beauce, Normandie, etc.), straw hats (Northern Italy), needles and nails (Normandy), swords (Northern Spain), watches and then plastics (Jura and Franche-Comté) among others, in the downtime of agricultural activity. Conversely, rural historians have not emphasised the strictly crafts aspect and have often presented these pluriactivities as a necessary response to the precariousness of life on a very small farm.

The question of reproduction and competitiveness of the small agricultural business in nineteenth- and early twentieth-century France was posed by Ronald Hubscher in an Annales article (Hubscher, 1985). For the first time this text offers a real conceptualisation of the notion of pluriactivity in rural life. Rejecting the idea that, until the early twentieth century, the success of a small farm can be explained precisely because it was supposed to have been primitive, self-sustaining, and marginal compared to the market economy, he proposes to study its capacity to adapt to the market in a micro-economic perspective, 'in terms of household economy rather than in terms of agricultural business economy.' The second activity, whether it concerned textiles, woodworking or metalworking, whether it was question of handicrafts at home or work in a factory, whether it rested upon the double activity of the head of the household or on the specialisation of tasks within the home, family or on the farm resulted in the farm's integration in the labour market and thus supplied it with ready money. In this way, the small business was tightly integrated into the overall economy, and it is tempting to interpret this practice of pluriactivity more as a strategy than a consequence of a weak production system that had to be endured. The small agricultural business's reproduction throughout the nineteenth century shows that pluriactivity were not only a response to the existing incompatibility between the size of a family and that of its farm, but that it was a part 'of a wider strategy intended to provide the household with regular flows of capital, sources of accumulated savings to create, increase, modernise, or else reconstruct its property.' Such an interpretation corresponds well to a conceptualisation pluriactivity in rural life, since it incorporates it in an overall social and economic environment. Pluriactivity was not only that of an individual, but of a family group. Neither was it limited to a farm, but implies taking the overall economy into account (Garrier, Hubsher, 1988). It could take the form of a double activity over time, especially in the mountains where rural activity was only possible in the summer and from where the men left for the plains to work as labourers or craftsmen during the winter (Simon, 2000). Abel Poitrineau's work, Remues d'hommes, describes the 'practical dualism of the mountain dweller, alternatively peasant and migrant worker' particularly well (Poitrineau, 1982: 144). It could also constitute a mixed activity throughout the year (Moulin, 1986) in the context of a 'putting-out system', or verlagsystem (Deyon and Mendels, 1981: 11-19). In its strictly rural and agricultural aspect, pluriactivity was complex. Although it was often a response to poverty, it sometimes led to undeniable economic successes, whether on a familial or regional scale (Antoine, 2004).

Pluriactivity was obviously not characteristic of all specialised farms just as all specialised farms did not practice pluriactivity. But a craft or proto-industrial activity practised by a farmer was most often integrated into a specialisation process in the context of a mixed productive system. It can be seen in this work in Gérard Le Bouedec's contribution (agriculture-fishing). The work of Alain Dewerpe on the formation of the 'industrial triangle' in Northern Italy during the nineteenth century can also be mentioned in support of this link between pluriactivity and farm specialisation (Dewerpe, 1985). At the origin of the process was the demographic overload in the countryside which required backup solutions to increase community revenue: 'Faced with an overpopulated countryside, rural communities resorted to multiplying crops which supplied the raw materials of proto-industry, made even easier because they then met with the desires of landlords interested in using them to increase their income.' Silk then became both a product and a raw material which enabled the increase in the resources of rural communities. During the eighteenth century, landlords, who were involved in this business developed a mixed system called fitto misto a grano: rent in grain for the land, rent in money for the house and a sharecropping contract (a *mezzadria*) for silk cocoons. The landlord then developed what Alain Dewerpe calls a 'worm strategy': he entrusted the sharecropper with a certain number of silkworms to raise, proportional to the quantity of blackberry leaves produced and the family's labour force. The system implied both a latent overpopulation in the countryside to raise the worms, and the development of silk weaving to occupy the population. Thus did it constitute a form of specialisation of agricultural activity developing in the context of pluriactivity.

At this stage of the reflection, we glimpse the possibility of also discussing, though in a different way, specialisation for large and small farms. Of course, it is recognised that large farms generally had greater aptitude to specialise (because they were bigger, they had more liquid assets, and so more means to make production choices requiring a starting investment, and they therefore had more chance of making a profit), but it is also known that small farms could perfectly well be profitable if...they specialised[27]. The visible paradox is that the aim was the same, only the means changed, depending on the farm's size.

Ultimately in analysing in detail the notion of pivotal production and that of the productive system, it has been shown that a large number of small or mid-sized businesses displayed at least a part of their activity which could be called specialised. The notion of economic success is ultimately included in the concept of specialisation. A specialised farm was one that, in its category, had the capability to both support the farmer and his family and to make a profit.

IV. Contributions of the neo-institutional movement and the notion of 'capability' to the issue of agricultural specialisation

The previous pages have explored the concept of specialisation. It has been observed several times that it could develop on small and large farms, even without further developing this difference. The mention of the neo-institutional movement in rural history will make it possible to go into depth on the agrarian systems' relationships to markets and institutional factors, and to envisage the modes of differential specialisation on different kinds of farms. It is then worth thinking about what the neo-institutional theory could bring to the question of specialisation in the agrarian systems of the early modern period and the nineteenth century. The contributions to this book are not all influenced by this movement. Nevertheless they do make it possible to look at TCE (Transaction Costs Economics) from three points of view. The first refers to the contracts that farmers signed with their landlords. The second refers to determining the institutions likely to favour agricultural growth, and to some extent the emergence of specialisation. The third, also fundamental, concerns the specificity of assets[28]. This point will make it possible, lastly, to question both the relationship of farmers to markets and the specialisation problem from a new angle, not reducible to the individual choices of farmers but on the contrary placed within a specific regional context.

Analysing the relationship between agriculture and markets has become obligatory since the work of Philip T. Hoffman on eighteenth-century French agriculture

[27] The many meanings of the term 'specialisation' appears very clearly in many works. For example, in the Barcelona region in the nineteenth century, Ramon Garrabou contrasts large farms dedicated to wheat and exclusively wine-producing small farms, and he notes that 'cereal sowing and wine growing were two different but complementary forms of agricultural specialization' Garrabou *et al.*, 2010: 150.

[28] 'Asset specificity has reference to the degree an asset can be redeployed to alternative uses and by alternative users without sacrifice of productive value.' (Williamson, 1991: 94). On the importance of this notion, see Joskow, 1991.

(Hoffman, 1996). In the absence of decisive technical progress, the transformations in this relationship seem to have been one of the driving elements in the process of agricultural growth in Western Europe between 1750 and 1940. If it is to be admitted that farm specialisation assumes a relationship to the market, the temptation to think about all the phenomena of agricultural specialisation with regard to the neo-institutional movement is very strong. O. E. Williamson indicated that 'Any issue that can be formulated as a contracting problem can be investigated to advantage in transaction cost economizing terms.' (Williamson, 1985: 17). From this point of view, the specialisation process which assumes a very high degree of dependence on the market should be able to be analysed in 'Williamsonian' terms. But in fact, this approach poses as many problems as it solves, particularly with regards to the modalities of specialisation of tiny tenant farms.

When he studied the problem of the company town, O. E. Williamson looked at the reasons why workers accepted agreements which were so evidently unfavourable to them[29]. The author sees two solutions. 'There are two leading possibilities. One is that students of company town have not performed the relevant comparative institutional test' (Williamson: 1985, 38). The second solution envisaged by O. E. Williamson is more unexpected. A chronic problem comes up in the organisation of the labour market: '…workers and their families are irrepressible optimists. They are taken in by vague assurances of good faith, by legally unenforceable promises, and by their own hopes for good life' (Williamson: 1985, 38). There is also a third explanation which brings up a fact which is difficult to ignore and makes it possible to temper the author's benign evaluation of the attitude of the labouring classes: the urgent need for workers to ensure their survival and that of their family[30]. This short digression makes it possible to contemplate one of the limits inherent to TCE.

In the same way that it is difficult in the context of TCE to explain the behaviour of workers in company towns, it is not very easy to explain why the micro-farms of Murcia accepted the unconscionable clauses written into the tenant contracts they concluded with the Counts del Valle. In this case, and in countless others, it seemed impossible to reduce the contracts between the farmers and the landlords, or those linking the peasants to their lords, to TCE patterns. If it is understood why the Count del Valle cared so much about shielding himself from the inherent risks in renting out his lands (which benefited from a complex irrigation system), this does not help to

[29] 'If contracting in its entirety obtains, then an efficient configuration of wages, home ownership, company store operations, and the like will appear, whatever the mobility condition of the population. What then explains the widespread discontent with the organization of company towns in the pre-mobility area?' (Williamson, 1985: 37).

[30] As Sheilagh Ogilvie notes, many historians when they study the efficiency of institutions, neglect the power plays that they hide (Ogilvie, 2007).

understand why small farmers accepted clauses so obviously not in their best interest. Maria Téreza Pérez-Picazo, sorely missed, is no longer here to answer this question. We must content ourselves with the two explanations advanced (above) by Oliver E. Williamson, although the validity of the second seems doubtful. Does that mean the neo-institutional approach must be entirely condemned? The answer is obviously negative, as Erich Landsteiner's contribution, showing the limits of the institution of servitude, given the specificities of wine-growing, clearly indicates[31].

Apart from Erich Landsteiner, several other contributors highlight the role of institutional changes in the specialisation processes. According to John Martin, government decisions in England aiming to ensure the nation's supplies during the Second World War provoked a de-specialisation of farms, resulting in a decrease in their efficiency. In Sweden on the other hand, the possibility of enclosure made it possible for farmers to increase their production. They turned more willingly to growing cereals, which were highly marketable. In Bulgaria, the collapse of the communist regime provoked dizzying growth in the number of farmers and a growth in self-sufficiency at first. After 2000, a concentration movement took on various forms.

In a slightly different vein, also greatly inspired by the neo-institutional movement, Pavel Matlas and Andreas Kuhlawy study the effects of the abolition and purchase of seigneurial rights on agricultural growth. From Bohemia to Brunswick, between the end of the eighteenth and the middle of the nineteenth centuries, feudal barriers fell. Depending on the place and the purchase conditions of seigneurial rights, the effect on agricultural production was more or less delayed and more or less favourable to the peasants. But ultimately, what comes out of these two contributions is the new institutional order that did not have an immediate effect in terms of specialisation. In Bohemia according to Pavel Matlas, the estates of the dukes of Schwarzenberg underwent a process of crop-intensification but did not become large specialised farms. As for Andreas Kuhlawy, he emphasizes the role of the State and the elites in the process of specialisation centred on sugar beet.

Lastly, though it is possible and even sometimes provable that institutional disruptions played a driving role in the specialisation (or de-specialisation) process, these different contributions demonstrate that models of regional agrarian development were subjected to constraints that reflect either the maintenance of medium- or long-term ecological balance, or the establishment of commercial and/or proto-industrial networks.

[31] It must be noted here that Erich Landsteiner indicates the most efficient contracts are those of salaried work and not those of tenant farming.

It is also worth examining the problem of the specificity of assets which is at the heart of O. E. Williamson's approach. It is with this concept's help that Williamson intends to analyse the phenomena of integration (or vertical disintegration). Indeed, one of the most important aspects of TCE refers to the question that can be formed in two complementary ways. Do the existing configurations of assets reflect technological principles? Is vertical integration the result of a natural technological order? Within the framework of TCE the answer to these two questions is clearly negative (Williamson, 1985: 85 and after).

In this context, the question must be whether the organisation of production processes of large farms, characterised by a strong degree of vertical integration[32], can be analysed in terms of investment specificity. To understand the point of this question, it must be split in two. On the one hand, we must decide whether the farms' assets were specific. On the other hand, the relevance of the concept in the context of traditional agriculture must be questioned, because as Williamson notes, its analysis finds its full legitimacy for the contemporary period[33]. From our point of view, the first question is the most important, and the second is greatly subordinate.

If it is admitted that a specific asset was not redeployable (see Williamson's definition cited above), and that this characteristic of assets was at the root of vertical integration phenomena, it must be asked what assets possessed this characteristic in traditional agricultures. This clarification will make it possible to understand better the vertical integration phenomena characteristic of the production process on large farms in particular.

From this point of view the work of Jean-Laurent Rosenthal on one hand, and Ingrid Henriksen et al. on the other are particularly important. Irrigation and drainage systems, or dairies, which Jean-Laurent Rosenthal and Ingrid Henriksen respectively consider, were clearly specific assets (Rosenthal, 1992; Henriksen, Hviid, Sharp, 2012). But were these not exceptions?

On large Belgian, German, or Czech farms, to only name a few examples from this volume, the essential input was produced on the farm itself. However it is not clear, except perhaps in the case of sugar refineries in the Duchy of Brunswick, how these agricultural businesses managed specific investments. What is more, for the latter it is worth highlighting that it was not the farm itself which constituted the specific

32 It is possible to speak of vertical integration as large farms were huge businesses which produced a large part of their input.

33 'I submit that the full range of organizational innovations that mark the development of the economic institutions of capitalism over the 150 past years warrant reassessment in transaction cost terms,' WILLIAMSON 1985: 17.

asset, but the sugar refinery. In the case of bankruptcy, as in the 1880s, the refinery did not find alternative employment, though the farms could turn toward other products. The refineries of the Duchy of Brunswick had a similar status to the Danish dairies mentioned in the work of Ingrid Henriksen.

Another approach is to explain vertical integration phenomena based on the idea of specificity of assets. The activity of farms conformed to the constraint of the product's geographic location and thus to one of the forms that the specificity of assets may take. However, the location constraint did not imply that the land, the livestock and the equipment had to be specific assets. Concerning the land, the dynamic markets (of land and leases) in a great many regions of Europe during the early modern and contemporary periods would suggest the contrary[34]. As for livestock and equipment, there arises a paradox. Either all assets of large farms will be considered highly specific (which would enable an explanation of the vertical integration processes put into place), or no asset will be considered specific. The second hypothesis seems to be the right one. For example, a dairy cow was likely to be transformed into hard cash on the second-hand market (she could be resold very easily, as shown by the abundance of such examples in the sources)[35].

There remains a last element of explanation which refers to the type of farm production. Wine, fruit, silk or olives, for example, were investments that were irreversible in the medium or perhaps even the long term in the sense that one could not remove the vines or trees and redeploy them as one could, say, with a wagons or livestock. But the fact that an investment was irreversible does not necessarily imply that it was a specific asset (in the sense of TCE). Erich Landsteiner's contribution in this volume shows that vineyards were not specific assets per se[36].

---

34 On the land market see Béaur, 1984, Boudjaaba, 2007; and for a recent clarification see Béaur, Chevet, 2013.

35 The same remark can be made for the different types of livestock (flocks of sheep, herds of cattle and horses, etc.) and equipment (plough, harrow, roller, etc.). Barns could be specific assets. But even in this case, it was not impossible to reinvest the elements of construction in other buildings, as shown by the dismantling of many buildings sold during the Revolution. Mills or large presses associated to certain large farms in the North of France could also be considered specific assets. But in no case could this characteristic alone explain the vertical integration choices of some large farmers. At this rate it will not be understood why certain millers had no farm, or inversely why most large farms had neither a press nor a mill. To understand these individual situations, there would have to be an *ad hoc* explanation for each farm and mill business which would assess the specificity or the non-specificity of the *mill* asset or the *press* asset.

36 'The easy access to land, the high labour intensity and small capital investments needed for cultivation, the practice of partible inheritance, a very lively land market and a low degree of social differentiation characterized viticultural societies in many regions.'

It is undeniable that vines were an irreversible investment. But, in the context of a given agrarian system, it was always possible sell one's vineyard and transform it into hard cash. It was only in the case of collapse and total liquidation of a regional wine industry that investments agreed upon by farmers and landlords became a total loss, meaning that they would have lost all their value[37].

Examples can be given of the complete disappearance of agro-economic systems which brought about a total evaporation of the values of investments agreed upon by the farmers (or the landlords). The Parisian vineyard, so essential to supplying Paris's wine in the eighteenth and early nineteenth centuries, completely disappeared between 1850 and 1880 due to the combined effects of powdery mildew, competition from Southern vineyards, and lastly, phylloxera. There are other French examples to mention: the slow agony of the silk industry in the Cevennes during the late nineteenth and early twentieth centuries, the brutal disappearance of the olive tree in certain regions of Southern France after the hard winter of 1709, etc. In these and many other cases, the agreed-upon investments really did lose all their value and were not likely to have alternative uses.

Lastly, in the context of the normal workings of traditional agrarian systems, the vertical integration phenomenon and that of specialisation around a pivotal crop, cannot be explained by the existence of specific assets. Incidentally Williamson quite rightly highlights that his analysis concerns the last one hundred and fifty years, or if the date of the first edition of his work is taken as a reference, a period roughly from 1830-1980. The chronological markers evidently owe nothing to chance.

In spite of the previous conclusions, there was a point to this long detour through TCE. It led to a distinction between two levels of analysis. The first refers to farmers' individual investment choices, the second to the structure of agrarian systems. Specialisation issues essentially played out at a regional level. They were rarely autonomous individual choices[38].

In order to understand this, the notion of 'capability' can once again serve as a guide. From the point of view of organisational economics, it represents the alternative to the explanation of vertical integration phenomena by the existence of specific assets. Sydney Winter, among others, emphasizes this concept to explain vertical integration (or disintegration) phenomena in industry or services (Winter, 1991; Jacobides, Winter, 2005). It is not question here of taking sides in a quarrel

[37] But what was true of wine as a product was not necessarily true of the vineyard's land.

[38] José Vicente Serrao illustrates this phenomenon with a very lovely example when he indicates the conditions of the emergence of wine specialisation in Madeira in the eighteenth century. Serrao, 2010: 54-56.

that goes far beyond the rural history field. It is simply a question of trying to explain what the term specialisation could mean for a farm in the early modern era and the nineteenth century, because this concept is applied indiscriminately to micro-farms hyper-specialised in their production, and to large farms involved in a range of activities which expanded as their size increased, but which at the same time evidently favoured one type of production to another.

In their 2005 article, Jacobides and Winter indicate on the one hand, that the specificity of assets was a subordinate phenomenon[39], and on the other hand that the phenomena of vertical integration and the extent of the firms' skills must be grasped at the industry level and not simply at the firm level. This second remark seems particularly fertile if cautiously reapplied to the rural history field[40].

It is indeed remarkable that the processes of farms conformed to geographical determinism, which is explained only in part by the commercial opportunities provided. The *Agricultural History of England and Wales* (volume V), for the period 1650-1750, illustrates this phenomenon with very great precision[41]. To respond to the observations of Winter and Jacobides, it would be worth reading the regional specialisations, which are presented according to farm size, as so many applications of specific skills whose specificities were determined by both the ecological environment and commercial opportunities. From this point of view, it is possible to understand the small degree of product-specialisation which characterised large farms, and to reinterpret the idea that peasant farmers would try to get the most out of their plot of land in order to reach the point of self-sufficiency.

Let us examine the first point, which is indisputably the most important as it makes it possible to better understand the dynamics of mixed farming. The de-specialisation of large farms should be interpreted as a growing phenomenon of integration and not as a phenomenon of horizontal diversification. Growing diversity of production should not be seen as a de-specialisation phenomenon in terms of the products, but as the effect of an environment and commercial opportunities. Depending on the period and the place, depending on the availability of labour and the ability of commercial networks to make the products more attractive, specialisation would then appear in the form of a pivotal crop which could be the sugar beet here, wheat there, livestock over there (for meat or milk production), barley in another place, etc. One of the most

39 The order of priority between these two concepts could easily be inverted. But it bears repeating that it is not a question here of taking sides.

40 Once again it must be clarified that it is not a question of slapping concepts onto particularly complex realities.

41 It is not worth multiplying examples to illustrate this phenomenon, which all contributions to this volume address.

remarkable examples of pivotal production making up the centre of gravity of large farms, without being the sole element, and based on specific skills, is discussed by Mohamed Ouerfelli. Sugar cane developed in the Mediterranean Basin throughout the Middle Ages, slowly progressing from East to West. At each step, planters had once again to solve the problems linked to the establishment of this crop, demanding of sun, water, wood, labour and know-how (Ouerfelli, 2008)[42]. As with sugar cane on large Mediterranean farms of the Middle Ages, the pivotal crops of large farms of the early modern era and the nineteenth century were not crops like the others. It was based on them, taking into account the environment, sometimes demography, and commercial opportunities, that the whole production process was organised.

In adopting this double point of view, it becomes possible to examine the distinct situations of large and small farms. Ecologically, small farms were less constrained that large ones, except in suffering the predations of powerful landowners. Of course, the environment wielded its priorities, but it would be easier to overcome the lack of fertilizer, or turn to a crop which used less fertilizer or water, or to specialise in a more labour-intensive plant if need be[43].

Thus the success of the potato among peasant farmers (in Flanders and countless other regions) and its absence from large cereal farms can easily be explained. This row crop used a lot of labour, and needed massive amounts of fertilizer. Lastly, its commercial value was doubtful. Although the crop met (for better but also for worse) the needs of a large percentage of the peasantry for nutritional security, it did not represent a profitable speculation on a large scale given the few commercial outlets and the costs in labour and fertilizer[44]. Other, very speculative crops such as flax, hemp, wine and dairy products[45] were symptomatic of these distinct specialisation phenomena.

We end here with a pattern which is rather far from what Jan de Vries proposed, in linking specialisation and the growth of scale of production. Ultimately, to show the effects of size of agricultural businesses on the level of specialisation, it would be worth setting aside specialisation in terms of products at the risk of hitting an

[42] The pages on Cyprus are particularly rich concerning ecological and demographic constraints. Those dedicated to Sicily show the sensitivity of this type of speculative crop to the economic situation and the risk of exhausting the soil. Lastly, the pages on Valencia show the pressing need to have much available labour. (Ouerfelli, 2008).

[43] The case of specialisation in olive growing by small Andalusian farms discussed by Juan Infante-Amate is particularly enlightening from this point of view (Infante-Amate, 2012).

[44] It is quite difficult to find an example of a French region where the potato became a large-scale object of speculation for large farms. The only exception would be the implantation of starch factories, which played a similar role to the sugar refineries and distilleries for sugar beet..

[45] In this volume see Pia Nilsson's contribution concerning hops in Sweden in the seventeenth century.

impasse: large farms would have been incapable of developing specialisations. On the other hand, by thinking in terms of *capabilities* and *pivotal crop*, it becomes possible to interpret distinct forms of specialisation by the different classes of farm in any given agrarian system.

The interest of this approach is not limited to what has been discussed above. It makes it possible to re-examine anew the small farm's aim of self-sufficiency. The progressive diversification of production, which accompanied the growth in cultivated area, refers not only to the peasant farmers' desire to obtain food independence. It also expresses an ecological necessity which would be more or less tempered by the availability of networks for marketing the different products.

Bibliography

ANSOFF, Igor (1957), 'Strategy for Diversification', *Harvard Business Review*, 35, 5, p. 113-124.

ANTOINE, Annie (1999), 'Systèmes agraires de la France de l'Ouest : une rationalité méconnue ?' *Revue d'Histoire Economique et Sociale*, 1999-1, p. 107-132.

ANTOINE, Annie (1994), *Fiefs et villages du Bas-Maine au XVIIIe siècle. Étude de la seigneurie et de la vie rurale,* Mayenne, Éditions Régionales de l'Ouest, Mayenne, 539 p.

ANTOINE, Annie, COCAUD, Martine (2004), 'La Pluriactivité dans les sociétés rurales. Approche historiographique', in Gérard LE BOUEDEC, François PLOUX, Christophe CÉRINO, Aliette GEISTDOERFER (eds), *Entre Terre et mer. Sociétés littorales et pluriactivité (XVe-XXe siècle),* Rennes, Presses Universitaires de Rennes, p. 13-33.

ANTOINE, Annie (2014), 'The Dubious Delights of Farm Accounts', in Gérard BÉAUR, Jean-Michel CHEVET (eds), *Measuring Agricultural Growth. Land and Labour Productivity in Western Europe from the Middle Ages to the Twentieth Century (England, France and Spain*, Turnhout, Brepols, CORN 15, p. 51-65.

BÉAUR, Gérard (2009), 'Alternative Agriculture or Agricultural Specialization in Early Modern France', *Agricultural History Review,* suppl. England-France A common agricultural heritage, p. 121-137.

BÉAUR, Gérard (1984), *Le Marché foncier à la veille de la Révolution. Les mouvements de propriété beaucerons dans les régions de Maintenon et de Janville de 1761 à 1790,* Paris, Éditions de l'EHESS, 360 p.

BÉAUR, Gérard, CHEVET, Jean-Michel (2013), 'Institutional Change and Agricultural Growth', in Gérard BÉAUR, Philipp SCHOFIELD, Jean-Michel CHEVET, Maria Teresa PEREZ-PICAZO (eds), *Property Rights, Land Markets and Economic Growth in the European Countryside*, Turnhout, Brepols, Rural History in Europe, 1, p. 19-68.

BEAUROY, Jacques (1998), 'Les Coke de Holkham Hall et l'essor du Norfolk system of Husbandry', *Histoire et Sociétés Rurales,* 10, p. 9-46.

BIAGIOLI, Giuliana (1989), 'Le Métayage en Italie centrale : un système agraire à l'épreuve de l'histoire et de l'historiographie', *Bulletin du Centre d'Histoire de la Région lyonnaise,* 3-4, p. 7-35.

BOUDJAABA, Fabrice (2008), *Des Paysans attachés à la terre ? Familles, marchés et patrimoines dans la région de Vernon (1750-1830)*, Paris, PUPS, 526 p.

BOURGUINAT, Nicolas (2002), *Les Grains du désordre, L'État face aux violences frumentaires dans la première moitié du XIXe siècle*, Paris, Éditions de l'EHESS, 544 p.

BOURRIGAUD, René (1994), *Le Développement agricole au 19e siècle en Loire-Atlantique,* Nantes, Centre d'histoire du travail de Nantes, 496 p.

BRAUDEL, Fernand (1979), *Civilisation matérielle, économie et capitalisme, XVe-XVIIIe siècle. Tome 3: Le temps du monde*, Paris, Armand Colin, 607 p.

BRAUDEL, Fernand (1984), *Civilization and Capitalism. 15th-18th century, The Perspective of the World*, Londres, Collins, 699 p.

BRUNT, Liam (2007), 'Where There's Muck, There's Brass: The Market for Manure in the Industrial Revolution', *Economic History Review*, 6, 2, p. 333-372.

CAMBIAIRE (DE), M. A. (1952), 'L'Autoconsommation agricole en France', *Bulletin de la Société française d'économie rurale,* 4, 3, p. 111-116.

CANTILLON, Richard (1755), *Essai sur la nature du commerce en général,* traduit de l'anglais, Londres, Fletcher Gyles, 430 p.

CLAR Ernesto, PINILLA, Vicente (2011), 'The Modernisation of Agriculture under Tough Environmental Constraints. One Hundred Years of Aragonese Agricultural Production, 1885-1985', in Mats OLSSON and Patrick SVENSSON (eds), *Growth and Stagnation in European Historical Agriculture*, Turnhout, Brepols, Rural History in Europe, 6, p. 195-228.

CONGOST, Rosa, JOVER, Gabriel, BIAGIOLI, Giuliana (eds) (2003), *L'Organització de l'espai rural a l'Europa Mediterrània : masos, possessions, poderi,* Girona, Curbet Comunicació Gràfica, Biblioteca d'historia rural, 380 p.

DE VRIES, Jan (1974), *The Dutch rural economy in the Golden Age, 1500-1700*, New Haven-London, Yale University Press, 316 p.

DELLEAUX, Fulgence (2012), *Les Censiers et les mutations des campagnes du Hainaut français. La formation originale d'une structure socio-économique (fin XVIIe-début du XIXe siècle),* Namur, Presses Universitaires de Namur, 230 p.

DELHOUME, Jean-Pierre (2009), *Les Campagnes limousines au XVIIIe siècle : une spécialisation bovine en pays de petite culture,* Limoges, Presses universitaires de Limoges, 455 p.

DERVILLE, Alain (1988), 'L'Assolement triennal dans la France du Nord au Moyen Âge', *Revue Historique,* 568, 112, p. 337-376.

DEWERPE, Alain (1985), *L'Industrie aux champs : essai sur la proto-industrialisation en Italie du Nord, 1800-1880,* Rome, École française de Rome, Paris, diffusion de Boccard, XXV-543 p.

DEYON, Pierre (1981), *La Proto-industrialisation : théorie et réalité,* Actes du colloque de Budapest, août 1982, *Revue du Nord,* 248, janvier-mars 1981, 308 p. See introduction by Pierre DEYON, Franklin MENDELS, p. 11-19.

DUBY, Georges (1962), *L'Économie rurale et la vie des campagnes dans l'Occident médiéval. France, Angleterre, Empire, IXe-XVe siècles : essai de synthèse et perspectives de recherches*, Paris, Aubier, 2 vol., 822 p.

FEDERICO, Giovanni (2005), *Feeding the World. An Economic History of Agriculture, 1800-2000*, Oxford, Princeton University Press, 388 p.

FONSECA, Helder Adegar, REIS, Jaime (2011), 'Limits of Agricultural Growth in a Fragile Eco-System. Total Factor Productivity in Alentejo, 1750-1850', in Mats OLSSON, Patrick SVENSSON (eds), *Growth and Stagnation in European Historical Agriculture*, Turnhout, Brepols, Rural History in Europe, 6, p. 142-167.

GARNIER, Bernard (1999), 'Des Bœufs pour Paris : commercialisation et élevage en Basse-Normandie (1700-1900)', *Annales de Bretagne et des Pays de l'Ouest,* 106, p. 101-120.

GARRABOU, Ramon, TELLO, Enric, CUSSO, Xavier, BADIA-MIRO, Marc (2010), 'Explaining Agrarian Specialization in an Advanced Organic Economy. The Province of Barcelona in the Mid-Nineteenth Century', in Vicente PINILLA (ed.), *Markets and Agricultural Change in Europe frome the 13th to the 20th century*, Turnhout, Brepols, Rural History in Europe, 2, p. 137-171.

GARRIER, Gilbert, HUBSCHER, Ronald (eds) (1988), *Entre Faucilles et marteaux : pluriactivité et stratégies paysannes,* Lyon, Presses universitaires de Lyon, Paris, éd. de la Maison des Sciences de l'Homme, 242 p.

GERVAIS, Michel, SERVOLIN, Claude, WEIL, Jean (1965), *Une France sans paysans*, Paris, Éditions du Seuil, 127 p.

GRÜNE, Niels (2007), 'Commerce and Community in the Countryside: The Social Ambiguity of Market-Oriented Farming in Pre-Industrial Northern South-West Germany (c.1770–1860)', *Rural History,* 18, 1, p. 71-93.

GURVIL, Clément (2011), 'La Pluriactivité des paysans parisiens au XVIe siècle: signe de précarité ou d'aisance?', *Mélanges de l'École française de Rome, Italie et Méditerranée*, 123, 1, p. 155-163.

GURVIL, Clément (2010), *Les Paysans de Paris du milieu du XVe siècle au début du XVIIe siècle,* Paris, Champion, 694 p.

Henriksen, Ingrid, O'Rourke, Kevin H. (2005), 'Incentives, Technology and the Shift to Year-Round Dairying in Late Nineteenth-Century Denmark', *Economic History Review*, 58, 3, p. 520-554.

Henriksen, Ingrid, Morten, Hviid, Sharp, Paul (2012), 'Law and peace: Contracts and the Success of the Danish Dairy Cooperatives', *Journal of Economic History*, 72, p. 197-224.

Herment, Laurent (2012), *Les Fruits du partage. Petits paysans du Bassin Parisien au xixe siècle,* Rennes, Presses Universitaires de Rennes, 395 p.

Herment, Laurent (2009), *Survivants ou Conquérants: Reproduction sociale et accumulation patrimoniale chez les petits exploitants agricoles de Seine-et-Oise durant le premier xixe siècle (1789-1860),* Unpublished Ph. D. EHESS, 2 vol., 1092 p.

Hoffman, Philipp T. (1996), *Growth in a Traditional Society: The French Countryside, 1450-1815*, Princeton, Princeton University Press, reprint 2000, 361 p.

Holderness, Brian Alexander (1984), 'East Anglia and the Fens/ Norfolk, Suffolk, *Cambridgeshire,* Ely, Huntingdonshire, Essex and the Lincolnshire Fens', in Joan Thirsk (ed.), *The Agrarian History of England and Wales. Regional farming systems,* V, 1, Cambridge University Press, p. 197-238.

Hubscher, Ronald (1985), 'La Petite exploitation en France : reproduction et compétitivité (fin xixe siècle-début xxe siècle)', *Annales, Economie, Sociétés, Civilisations,* 1985-1, p. 3-34.

Huriot, Jean-Marie (1994), *Von Thünen. Economie et espace, Paris,* Economica, 352 p.

Infante-Amate, Juan (2012), 'The Ecology and History of the Mediterranean Olive Grove: The Spanish Great Expansion, 1750-2000', *Rural History*, 23, 2, p. 161-184.

Jacobides, Michael G., Winter, Sidney G. (2005), 'The Co-Evolution of Capabilities and Transaction Costs: Explaining the Institutional Structure of Production', *Strategic Management Journal*, 26, p. 395-413.

Joskow, Paul. L. (1991), 'Asset Specificity and the Structure of Vertical Relationship: Empirical Evidence', in Oliver E. Williamson, Sidney G. Winter (eds), *The Nature of the Firm. Origins, Evolution, and Development*, Oxford, Oxford University Press, p. 117-137.

Kim, Kyung-Keun (1998 et 1999), 'Le Vexin français de 1700 à 1850 : une économie rurale préindustrielle en progrès », with Jacques Dupâquier's presentation, *Mémoires de la Société Historique et Archéologique de Pontoise, du Val d'Oise et du Vexin*, 36, 1998 et 37, 1999, 94 p. and 70 p.

Knibbe, Merijn (2011), 'Agricultural Productivity the Coastal and Inlandarea of Friesland, 1700-1850', in Mats Olsson, Patrick Svensson (eds), *Growth and Stagnation in European Historical Agriculture*, Turnhout, Brepols, Rural History in Europe, 6, p. 83-116.

Konersmann, Frank (2011), 'Land and Labour Intensification in the Agricultural Modernization of Southwest Germany, 1760-1860', in Mats Olsson, Patrick Svensson (eds), *Growth and Stagnation in European Historical Agriculture*, Turnhout, Brepols, Rural History in Europe, 6, p. 141-166.

KOPSIDIS, Michael (2010), "Boserup meets Thuenen: Markets and farming intensity in pre-industrial Westphalian peasant agriculture, *circa* 1830", in Vicente PINILLA (ed.), *Markets and Agricultural Change in Europe from the 13th to the 20th century*, Turnhout, Brepols, Rural History in Europe, 2, p. 107-135.

KOPSIDIS, Michael, WOLF, Nikolaus (2010), 'Agricultural Productivity Across Prussia During the Industrial Revolution: A Thünen Perspective,' *The Journal of Economic History*, Cambridge University Press, 72, 3, p. 634-670.

LEMARCHAND, Yannick (1993), *Du Dépérissement à l'amortissement, enquête sur l'histoire d'un concept et de sa traduction comptable*, Quimper, Ouest Éditions, 719 p.

MARGAIRAZ, Dominique (1988), *Foires et marchés dans la France préindustrielle,* Paris, Éditions de l'EHESS, 275 p.

MEUVRET, Jean (1988), *Le Problème des subsistances à l'époque de Louis XIV. Tome III : Le commerce des grains et la conjoncture,* Paris, Éditions de l'EHESS, 2 vol., 191 et 162 p.

MEUVRET, Jean (1953), 'Agronomie et jardinage aux XVIe et XVIIe siècles', in *Éventail de l'histoire vivante. Hommage à Lucien Febvre,* Paris, A. Colin, II, p. 353-362.

MORICEAU, Jean-Marc (1994), *Les Fermiers de l'Île-de-France : l'ascension d'un patronat agricole (XVe-XVIIIe siècle),* Paris, Fayard, 1069 p.

MORICEAU, Jean-Marc, POSTEL-VINAY, Gilles (1992), *Ferme entreprise famille : grande exploitation et changements agricoles : les Chartier XVIIe-XIXe siècles*, Paris, Éditions de l'EHESS, 397 p.

MOSER, Peter, VARLEY, Tony (2013), 'The State and Agricultural Modernisation in the Nineteenth and Twentieth Centuries in Europe', in Peter MOSER, Tony VARLEY (eds), *Integration through Subordination. The Politics of Agricultural Modernisation in Industrial Europe*, Turnhout, Brepols, Rural History in Europe, 8, p. 13-39.

MOULIN, Annie (1986), *Les Maçons de la Creuse, les origines du mouvement,* Clermont-Ferrand, Publication de l'Institut d'Études du Massif Central, rééd. 1994, X-564 p.

MULLIEZ, Jacques (1979), 'Du Blé, «mal nécessaire». Réflexions sur les progrès de l'agriculture de 1750 à 1850', *Revue d'Histoire Moderne et Contemporaine*, 26, 1, p. 3-47.

OGILVIE, Sheilag (2007), ''Whatever Is, Is Right'? Economic Institutions in Pre-Industrial Europe', *Economic History Review,* 60, 4, p. 649-684.

OLMSTEAD, Alan L., RHODE, Paul (1993), 'Induced Innovation in American Agriculture: A Reconsideration', *Journal of Political Economy,* 101, 1, p. 100-118.

OLSSON, Mats, SVENSSON, Patrick (2011), 'Agricultural Production in Southern Sweden, 1702-1864', in Mats OLSSON, Patrick SVENSSON, (eds), *Growth and stagnation in European historical agriculture*, Turnhout, Brepols, Rural History in Europe, 6, p. 117-140.

OUERFELLI, Mohamed (2008), *Le Sucre. Production, commercialisation et usages dans la Méditerranée médiévale*, Leyden-Boston, Brill, 809 p.

PINILLA, Vicente (2009), 'The Impact of Markets in the Management Rural Land', in Vicente PINILLA (ed.), *Markets and Agricultural Change in Europe frome the 13th to the 20th century,* Turnhout, Brepols, Rural History in Europe, 2, p. 11-36.

PEREN, Richard (1989), 'Market and Marketing', in Gordon E. MINGAY (ed.), *The Agrarian history of England and Wales, VI, 1750-1850*, Cambridge University Press, p. 190-274.

POITRINEAU, Abel (1982), *Remues d'hommes : essai sur les migrations montagnardes en France aux XVIIe et XVIIIe siècles,* Paris, Aubier Montaigne, coll. « Collection historique », 325 p.

POURCHASSE, Pierrick (2006), *Le Commerce du Nord : les échanges commerciaux entre la France et l'Europe septentrionale au XVIIIe siècle,* Rennes, Presses Universitaires de Rennes, coll. « Histoire », 392 p.

POSTEL-VINAY, Gilles (1974), *La Rente foncière dans le capitalisme agricole,* Paris, Maspéro, 286 p.

POUSSOU, Jean-Pierre (1999), ''*L'Agriculture alternative*' : à propos d'un livre de Joan Thirsk', *Histoire et Sociétés Rurales,* 12, 2, p. 131-147.

RAMANUJAM, Vasudevan, Varadarajan, P. (1989), 'Research on Corporate Diversification: A Synthesis', *Statregic Management Journal*, 10, 6, p. 523-551.

RICHARDSON, George B. (1972), 'The Organisation of Industry', *Journal of Economic Literature,* 82, 327, p. 883-896.

RONSIJN, Wouter (2013), 'Mediated and Unmediated Market Dependence in the Flemish Countryside in the Nineteenth Century: Two Paths of Rural Development', *Continuity and Change,* 28, 1, p. 89-120.

ROSENTHAL, Jean-Laurent (1992), *The Fruits of Revolution, Property Rights, Litigation and French Agriculture (1700-1860),* Cambridge, New York, Melbourne, Cambridge University Press, 216 p.

SAGUER, Enric (ed.) (2011), *Forçats a foc i llum: Una història oral dels ultims masovers de la regió de Girona, 1930-2000*, Barcelona (ESP), Generalitat de Catalunya, Departament de Cultura.

SCOLA, Roger (1992), *Feeding the Victorian City: The Food Supply of Manchester, 1770-1870*, Manchester, New York, Manchester University Press, Manchester, xx-347 p.

SERVOLIN, Claude (1972), 'L'Absorption de l'agriculture dans le mode de production capitaliste', *L'Univers politique des paysans,* Paris, Armand Colin, p. 41-77.

SERRAO, José Vicente (2010), 'Land Management Responses to Market Changes. Portugal, Seventeenth-Ninetennth Centuries', in Vicente PINILLA (ed.), *Markets and Agricultural Change in Europe frome the 13th to the 20th century*, Turnhout, Brepols, Rural History in Europe, 2, p. 47-73.

SIMON, Anthony (2000), *La Pluriactivité dans l'agriculture des montagnes françaises: étude géographique,* Ph. D., University Blaise Pascal (Clermont-Ferrand), 568 p.

THIRSK, Joan, *(1997), Alternative Agriculture. A History, From the Black Death to the present Day*, Oxford, New York, Athènes, Oxford University Press, 365 p.

THIRSK, Joan (1992), 'L'Agriculture et les plantes nouvelles en Angleterre aux XVIe et XVIIe siècles', in *Plantes et Cultures nouvelles en Europe occidentale au Moyen Âge et à l'époque moderne. Actes des Douzièmes Journées internationales d'histoire du centre culturel de l'abbaye de Flaran,* Centre culturel de l'Abbaye de Flaran, Auch, p. 69-80.

WARDE, Paul (2006), 'Subsistence and Sales: The Peasant Economy of Württemberg in the Early Seventeenth Century', *Economic History Review,* LIX, 2, p. 289-319.

WILLIAMSON, Oliver E. (1991), 'The Logic of Economic Organization', in Olivier E. WILLIAMSON, Sidney G. WINTER (eds), *The Nature of the Firm. Origins, Evolution, and Development*, Oxford, Oxford University Press, p. 90-116,

WILLIAMSON, Oliver E. (1985), *The Economic Institutions of Capitalism,* New York, The Free Press, London, Macmillan, 450 p.

WINTER, Sidney G, 'On Coase, Competence, and the Corporation', in Olivier E. WILLIAMSON, Sidney G. WINTER (eds), *The Nature of the Firm. Origins, Evolution, and Development*, Oxford, Oxford University Press, p. 179-195.

I

Specialisation: the result of farming skills

2. Early specialisations and late specialisations in the rural area around Liège: an evaluation at the beginning of the twentieth century

Paul SERVAIS

I. Introduction

Without understanding specialisation, it is impossible to comprehend the modernization process that transformed the European countryside in the period from early modern times to the present day. Specialisation, as discussed by Franklin Mendels in his model of proto-industrial development (Mendels, 1972), refers back to the theory of comparative advantage. It requires us to look not only for its origin and the necessary conditions for its development, but also at its very nature: does specialisation concern a particular type of production, such as cereals, cattle breeding, vine growing, industrial crops and so on? Or does it refer to the choice of a specific kind of farming: intensive as opposed to extensive cultivation, massive capital investment or heavy labour input; smallholding (truck-farming for example) in contrast to large-scale farming enterprises like single-crop cereal farming?

The countryside around the city of Liège witnessed various attempts at agricultural specialisation between the eighteenth and the twentieth centuries. It is therefore possible to follow the different stages of the specialisation process and to analyze its causes and results not only at the level of the sub-region but also on the farm. Drawing on various surveys from the period between 1800 and 1914, along with farm accounts and administrative records, this article focuses on the chronology, geography and typology of agricultural specialisation in the Liège countryside during this period. Its purpose is to sketch briefly the factors that started the process and kept it going; to see how it was carried out, the effect it had on the level of farming, and its impact on the standard of living and on agricultural structures.

Agricultural specialisation and rural patterns of development, ed. by Annie ANTOINE, Turnhout, 2016 (*Rural History in Europe*, 12), p. 49-62.

BREPOLS PUBLISHERS DOI 10.1484/M.RURHE-EB.5.112260

II. Forms of specialisation in the Liège countryside

Over the past three centuries the process of general modernization in the countryside around Liège produced three specific regional types of agricultural specialisation. The first signs of this general modernization appeared as early as the sixteenth century, and its rhythm picked up between the late eighteenth and through the nineteenth century before breaking down in the 1950s.

Figure 2.1. The Liège region

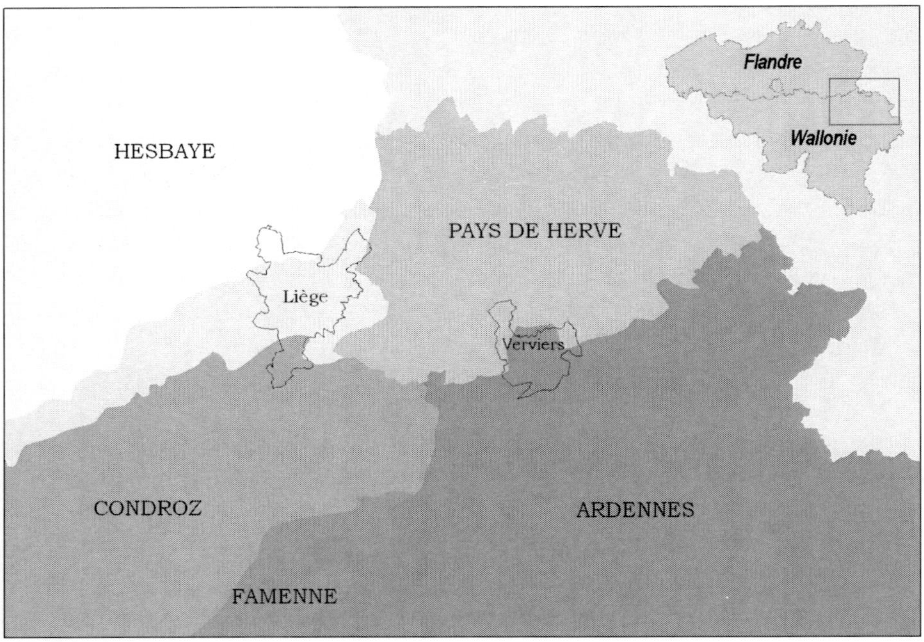

Source. Map by the author.

The first form of specialisation, visible as early as the 1550s (Ruwet, 1942) in the area known as the Pays de Herve, was characterized by small-scale dairy farming and fruit production. There was also a system of small holdings. From the 1820s and 1830s onward, this agriculture attracted a good deal of middle-class investment and rapidly undermined the structures of proto-industry established during the eighteenth century (Servais, 1982a, 1982b and 1982c). It was well integrated in the circuits of international trade, which made credible its claim to be 'the most modern countryside in Europe'. This was further reinforced by the development of food production on an industrial scale: first seen with cheese, then with fruit syrups and cider, all of which were widely marketed (Bihot, 1913).

A second form of specialisation appeared as early as the seventeenth and eighteenth centuries in the area of Hesbaye. Here Church and noble land predominated at first, to be supplanted in the nineteenth century by institutional and middle-class property (Delatte, 1945: 71-78; Brulard, 1962). Hesbaye was dominated by large cereal farms, producing for nearby town markets. From the mid-nineteenth century on, industrialization proceeded apace, as the sugar-beet came into cultivation and infrastructures grew up specifically for its processing (Vliebergh and Ulens, 1907; Neven and Oris, 2004).

At some distance from these lay a third region, the Ardennes, characterized until the nineteenth century by a mix of small holdings with some big Church and noble estates, smaller tracts of middle-class property, and vast amounts of communal land. Agriculture was very traditional here, primarily focused on simple household subsistence. Specialisation took place only very late, during the last third of the nineteenth century, when the regional economy turned to dairy farming, meat production, horse-breeding and forestry (Delatte, 1945: 92-95; Hoyois, 1949-1953; Vliebergh & Ulens, 1911).

Overall, these trends towards specialisation were encouraged by two major forces. First, the region had active local, regional and even international markets. Secondly, it had, or created, circulation and transportation networks. Market structures resulted above all from a relatively high population density and an extensive urban network. The circulation and transportation networks were already in operation for the most part during the early modern period and in any event underwent major changes during the nineteenth century, as mobility increased and still further attempts were made to open up the region.

Other conditions may have encouraged the process of specialisation in specific ways: the increasing use of money in the economy, the availability of capital for investment, the models of agricultural development that were promoted, and complementary non-agricultural activities, which could ease the transition from mixed farming to specialisation and the development of regional complementarities.

The drive to specialisation seems mainly to have been political, though that could work positively or negatively. The oldest of these regional specialisations, dairy farming in the Herve region, started when the borders of the Spanish Netherlands were closed to cereal exports and the territory was wedged between the duchy of Limbourg and the county of Dalhem (Ruwet, 1942: 179-86). In the final stage, in the massif of the Ardennes, political decisions, particularly the 1847 law on clearings, were also a determining factor, though the changes were partly a result of local culture, and partly inspired by the example of other nearby regions, particularly the Pays de Herve.

III. Impact on farming

These moves to specialisation, each with its particular characteristics and specific chronology, had an important effect on labour organization on the farm, on the capital mobilized for operation, total farm budget, average size, and the standard of living and priorities of the farmers. What changes occurred by the end of the nineteenth and the beginning of the twentieth century?

III.1. The Herve region and the family farm

After four centuries of change, and a period of rapid industrialization, the Pays de Herve had lost its non-agricultural rural industries. This zone had in fact been ruralised from a very early date and specialized in dairy production. The standard unit was basically the family farm; as Charles Bihot noted, in 1913, 'a family of peasants, can, with two workers in addition to the father and the mother, take on a medium-sized holding on its own...'

Farm units operated under a strict specialisation of labour along gender lines. The farmer did the wholly outdoor work: he bought and sold cattle, and carefully maintained the meadows by eradicating parasitic plants, spreading dung, exterminating moles and moss, hauling manure, repairing fences, pruning hedges, transporting produce, mowing and tending the hay. But it was the woman who was at the heart of the production process. In addition to supervising the farmyard, she made butter and cheese, gathered fruits, prepared syrup, cleaned the dairy utensils and workrooms and kept whatever family accounts there were. All the members of the family participated in milking, however, which was an essential task, repeated thrice daily.

In addition to this strict allocation of tasks, the growth of mechanization, requiring more know-how and better process control, made the workplace more specialised; it was divided and ordered as strictly as possible into spaces dedicated to each successive stage of operations. This was a true production line, its efficiency and the increase in production were beyond doubt.

Although there are no farm accounts from the end of nineteenth or the beginning of the twentieth century, agricultural surveys (the *Statistique de la Belgique*, 1895) and

'Agricultural Monographs', together with local agricultural statistics and inquiries, make it possible to estimate a farm budget.

Farm Income. The butter yield of a cow can be reckoned at 300 francs/year, more or less. If the milk was used instead to produce cheese, we can estimate an output of 800 francs/year for each cow. This is based on an average production of 4,000 litres of milk per year; one piece of the best quality cheese used up ten litres of milk and sold for two francs. The income of a small-scale farming unit of around five hectares, sufficient to feed ten cows, thus came to between 3,000 and 8,000 francs a year, more probably around 5,500 francs, if the milk produced was allotted in the usual fashion between butter and the various types of cheese. Depending on what stage in the life-cycle the household had reached, producers might have to set aside 500 francs/year to pay for their servants, at least if no family members could assist the farming couple. Operating profits and farm revenue can thus be estimated at approximately 5,000 francs/year.

Farm expenses. The cost of one hectare of meadow varied between 5,000 and 9,000 francs; the purchase price of a milk cow could run as high as 700 francs; the total working assets required for one hectare can be reckoned at 2,000 francs. A maximum estimate of investment on this five-hectare farm would therefore be: land investment: 45,000 francs; cattle purchases: 7,000 francs; working assets: 10,000 francs. This comes to a total investment of 62,000 francs. At a 3 per cent mortgage rate, this implies an annual outlay of 1,860 francs. If the land was free of all charges – a very rare situation according to our studies of family transmission practices and strategies – the yearly outlay might be as low as 510 francs. It is harder to estimate the cost of household labour and household consumption. According to oral history surveys, self-sufficiency on these items remained the ideal down to the 1950s. Thus net household income, not including everyday family expenses, can be evaluated around 3,140 francs or 4,490 francs in case of land free of charges.

For the farmer of this five-hectare farm, the *return on investment and on household labour* for the working assets and livestock alone, would have come to 18.47 per cent or, in case of free land, 26.41 per cent. For the total invested capital it stood at 5.06 or 7.24 per cent. However it should be stressed that these returns are minima: they do not take into account additional income from the artisanal *siroperies*, cider-houses, and distilling or canning facilities which were found on almost all the farms. Nor do they include income from the farmyard or the pigsty, nor sales of entire crops to industrial food companies, which grew during the last quarter of the nineteenth century.

During the eighteenth century, the structure of invested capital was slightly different, as working assets were not mentioned as such in the records. For a farm of four *bonniers*[1], representing a capital of 3,200 guilders, with livestock consisting of six cows worth 690 guilders in all, gross income can be estimated at 425 guilders and the mortgage payment at 194.5 guilders. The minimum net income of the farm and the household can be evaluated at 230.5 guilders. Daily consumption and daily household expenses are just as difficult to evaluate as in the later period. Never the less the return on investment and household labour can be estimated at, respectively, 33.4 per cent for livestock and 5.9 per cent for total capital. If the land was free of mortgage charges, minimum net income and gross income would have been identical. In this case, the return on investment and labour can be evaluated at 61.59 per cent for the livestock and 10.9 percent for total capital. Over a hundred years, the regression of the return from livestock and working assets are closely correlated, as is that of overall investment, except in the case of land not bearing charges.

All in all, we can say that at its outset this early form of specialisation freed up a good deal of labour, permitting farmers to concentrate on activities with high added value and on developing non-agricultural domestic activities. During a second stage, processing techniques were improved, in particular with gradual mechanization; this made the production process increasingly effective. This change worked against the de-industrialization of the countryside, substituting new industrial food production activities for the domestic industries of the eighteenth and early nineteenth centuries. It helped integrate the farming system in local, regional and international markets, and encouraged greater investment in it, particularly in machinery. This last point may explain the lower rate of gross profits at the end of the nineteenth century compared to the rate of gross profit at the end of the eighteenth century. But the amount of income was itself increasing proportionately to the growth of production. This very modern situation was probably why transmission of property within the family was used to keep holdings intact (Servais, 2003; Servais, 2005). It may explain why the property market was livelier in this area than in the other two regions (Servais, 1994).

III.2. Hesbaye and the large farm

Before the nineteenth century the Hesbaye region concentrated primarily on cereal production. But when the sugar beet was introduced in the second half of the nineteenth century, it set off a farming revolution and expanded the use of money in the economy. The range of agricultural productions shrank drastically.

In 1828, land on a large farmstead of around ninety hectares was allotted to different usages as follows:

[1] Around 3.5 hectares.

Table 2.1. Land use on a large farm in 1828

Crops	*percentage of area in 1828*
wheat	12.0
rye	15.0
German wheat	3.0
barley	2.0
oats	15.0
clover	10.0
fodder	10.0
colza	2.5
potatoes	1.5
carrots	1.0
fallow	20.0
natural meadows	6.0
buildings	2.0
	100

In 1907, this arrangement was simplified:

Table 2.2. Land use on a large farm in 1907

Crops	*percentage of area in 1907*
sugar beets	20.0
mangel-wurzels	1.0
oats	30.0
wheat	180.
clover	5.0
permanent meadows and orchards	15.0
temporary grazing	10.0
buildings	2.0
	100.0

Source. VLIEBERGH & ULENS, 1907: 90.

The choice of cereal grown resulted less from scientific and regular rotation than it did from market trends, particularly so with the sugar-beet. Sugar manufacturers, who might also be manure-merchants, delivered fertiliser on credit to the small and medium farms, advanced necessary cash and, finally, settled for the beets in a single payment, which came to a large sum. Farmers on large farms were also sensitive to these pressures and were aware that, as maintaining the beet fields required much work, the sugar-beet kept all their servants fully employed. Leaves and pulp, the by-products of sugar-beet cultivation, also made this crop attractive, as they provided abundant cattle fodder. Cattle breeding was without doubt a major household activity. A farm of around sixty hectares fed nearly eighty oxen or cows raised for their meat, and incidentally, their milk. These cattle were entrusted to the farmer by cattle merchants and fed on the by-products of cereal or beet cultivation, sometimes supplemented with by-products bought from dairy farms in the Pays de Herve; they ended their career on the wholesale meat-markets of Liège and Anderlecht. The breeding of draught horses, and pig raising, on a very large scale, might also be included in this category. By contrast, sheep-rearing, which had been very common before the beginning of the nineteenth century, now practically disappeared (Brulard, 1962).

Unlike farming in the Pays de Herve, the process here – at least on large and medium farms – depended less on the labour of the farm family than on salaried hands. Thus, an observer notes:

The large farmer directs his farm and supervises it. His wife takes care of the farmyard and sometimes the cattle-shed. But on large farms now, young women tend to neglect the cattle and even the farmyard. In a way this has an advantage: they can take better care of the household and children, and make their husband's life more comfortable (Vliebergh & Ulens, 1907: 86).

An intensively worked seventy-hectare farm in 1907, with 12 hectares of beets, 14 of wheat, 18 of oats, 2.6 of rye, 2.6 of barley, 0.4 of potatoes, 8 of clover and 11 of meadows (Vliebergh & Ulens, 1907: 91), provided jobs for thirteen workmen, plus a foreman and a maidservant. The wages averaged around 1.75 francs a day over 300 working days. These were winter wages; they were higher during the summer, at 2 francs, and reached 2.50 francs a day 'during the six weeks of the harvest'. Various supplements or regular bonuses augmented this salary, depending on the worker's age, gender or expertise.

The farm at the Trognée château had 25 hectares under sugar beets, 25 in oats, 23 in wheat, 2 in rye, 3 in winter barley (escourgeon), 3 under potatoes and 10 in red clover; it also had 10 hectares of meadows. It employed twenty-nine people, including ten temporary day workers, used more particularly 'at sugar beet time (for hoeing, thinning and pulling up the crop), for the hay harvest, the grain harvest, for setting and pulling up potatoes, for threshing grain with the steam engine'. It should be noted that this farm had a significant number of cattle, both for fattening (40 oxen from May to September) and for work (10 to 12 mares and plough geldings, 6 to 7 plough foals, 7 to 8 draught oxen, 5 milk cows, 2 heifers and 2 calves), which required particular care. Even greater care was taken of the three thoroughbred horses mentioned in the survey – a clear indication of the high standard of living enjoyed by these 'gentleman' farmers (Vliebergh & Ulens, 1907: 299-305).

The size of these farm units naturally raises the question of the working assets, that is, the sum needed by a farmer without animals or tools, in order to take on a farm. Theoretical evaluations made at the beginning of the twentieth century, and the end of the process of specialisation, indeed the industrialization, of farming, vary from 400 to 1,000 francs per hectare. This sum covers not only wagons and carts, harrows and ploughs, but also, and especially, livestock. For a farm of one hundred hectares, this amount came to at least 80,000 gold francs, which necessarily implies a recourse to credit.

Operating accounts, in the rare cases when they can be found, make it possible to reconstitute the broad outlines of farm budgets and gross income, as well as the funded capital (exclusive of the amount required to purchase the land).

For a small-scale farm of five hectares, held on a lease, the results are as follows. Expenses included pulp (108 gold francs), bran (170 fr.), oil-cakes (50 fr.), manure (140 fr.), seeds (20 fr.), piglets (100 fr.), rent (900 fr.) and repairs (75 fr.) for a total of 1,563 gold francs. Revenues came from the sale of beets (875 fr.), oats (570 fr.), potatoes (50 fr.), pigs (500 fr.), cattle (400 fr.), butter (740 fr.) and eggs (120 fr.), which added up to 3,255 gold francs. Gross income of the farm household was 1,692 francs; to this should be added the income from wheat produced as a subsistence crop on the farm, estimated at 160 francs. For a working asset of 2,500 francs, this comes to a total of 1,852 francs, leaving aside the estimated or theoretical value of labour by the father, the son and the mother. The return on investments and household labour can thus be estimated at 74.08 per cent (Vliebergh & Ulens, 1907: 103). A farm with exactly the same characteristics,but located a little farther to the south of Hesbaye produced a return of 80 percent on its working asset and family labour.

These may be compared with accounts for a hundred-hectare farm in the same area. Expenses (all in gold francs), included an annual rent of 14,000 francs, cost of seeds and manure (15,000 fr.), wages (5,250 fr.), losses reckoned at 2,000 francs, and lastly the purchase of pulps, oil cakes and the like (5,000 fr.); in all, a total of 41,250 francs. Income from crops on 30 ha of beets, 50 of cereals and 2 of potatoes came to 44,200 francs (beets, 20,700 fr.; cereals, 22,500 fr.; potatoes, 1,000 fr.) The farm earned 2,500 francs from livestock: fattening 25 oxen brought in 1,250 fr; the sale of 5 foals and 5 calves, 2,000 and 500 fr respectively, for a total of 2,500 fr; farmyard and dairy sales accounted for about 2,000 additional francs. Total gross income totalled 49,950 fr. (Vliebergh and Ulens, 1907: 342-343). Net income thus amounted to 8,700 francs: in other words, 15.81 per cent on a working asset of 55,000 francs plus additional household labour. What about the return on land? At a rate of 5,000 francs/hectare, the 100-hectare farm was worth around 500,000 fr. The tenant's rent of 14,000 francs did not exceed a return of 2.9 per cent. On the five-hectare farm, worth 25,000 fr. more or less, the rent would have been 900 francs, producing a return of 3.6 per cent. For a landowner, the return on the total investment would be around 3.54 per cent on the 100-hectare farm and 4.55 per cent on the five-hectare farm.

The difference between the output of the working asset and that from landed capital could explain at least in part why the farm in Hesbaye was usually let to a tenant, and not worked by the owner. The high amount of capital needed at the same time, to cover operating and the purchase of tools seems to have made land investment, in a far from active market, very difficult (Servais, 1994). This may explain why the transmission of family assets was mainly about passing on leases to the next generation (Servais, 2003; Servais, 2005). The income earned from the working asset also makes it possible to understand the initially gradual and subsequently intensive orientation of the large farm towards the market. Small-scale holdings survived only if the tiny share of land or built-up property they contained was enough to allow an agricultural labourer the hope of living of his own, free of the need to sell his labour.

The better financial performance of small-scale farming can undoubtedly be attributed to greater input of family labour and a lower volume of investment. In terms of the standard of living, the gross income of the large farm made the holding an autonomous unit and allowed the farmer to live comfortably; this is evident from the succession declarations and probate inventories, and can also be seen in the history of the families involved. Compared to the situation in the eighteenth century, the budgets of all farm categories had clearly improved in proportion to their size. But, as in the Pays de Herve, the level of return was declining, because farm operation required increasing amounts of capital.

III.3. Ardennes and the small family property

At the beginning of the twentieth century, the typical Ardennes farm was similar to the model seen in the Pays de Herve, but adapted to the harsher conditions of the Ardennes massif where soil quality was poorer. The 'small' farm was bigger here, around twenty hectares in area. Farms focused on breeding dairy stock, oxen and pigs for meat and a huge number of horses. They also produced great quantities of oats for local consumption, and for markets outside the region. Of course production varied according to the size of the farm.

The cost of land and working assets rose significantly between 1870 and 1914. This was particularly the case here because heavy emigration made labour scarce and mechanization, increasingly necessary. This pushed up production costs. Around 1870, working assets came to roughly 200 francs per hectare; in 1911, their cost had reached about 700 francs. Working assets consisted predominantly of livestock, though the amounts varied according to the size of the farm.

For a medium sized farm (52 ha.), it was made up as follows (all values in gold francs): horses, 4,200 fr. ; oxen, pigs and sheep: 12,545 fr.; hens, pullets, chicken and ducks: 140 fr.; dogs: 50 fr.; tools and equipment: 3,488 fr. All the elements together came to 20,423 fr., or 392 fr./ha. For the same farm, soil use was as follows: permanent meadows, 21 ha.; temporary meadows, 14 ha.; spring cereals, 11 ha.; rye, 2 ha.; potatoes, 2 ha.; root crops, 2 ha. (Vliebergh & Ulens, 1911: 119).

On this basis, we can estimate expenses and income as follows (all values in gold francs). Expenditures : annual rent, 3,120 fr., wages, around 2,400 fr.; fertilizer and livestock fodder, 10,400 fr, for a total of 15,920 fr. Income: livestock sales: 5.400 fr. (horses, 1,600 fr., cows, 1,800 fr.; sheep and pigs, 2,000 fr.) and dairy sales: 13,125 fr., for a total of 18,525 fr.

Thus total net farm revenue was around 2,605 francs, or 50.01 francs per hectare. The return on the working assets and family labour was a little over 10 per cent, clearly less than the average in the two other regions. The return on total investment was much the same. Measured at market prices, the overall value of this fifty- two hectare farm can be reckoned at around 150,000 francs. Total investment therefore came to about 170,000 francs and the return on this total investment was no higher than 1.53 per cent, clearly less than in the Pays de Herve and the Hesbaye area, even if the farm was not burdened by mortgage charges or owned by the farmer himself, and even though families worked even harder there.

The Ardennes region did not stand in total contrast to the two other regions, particularly as regards its standard of living. But its economic situation of the region had clearly improved in comparison to what it had been in the eighteenth century. Large tracts of common forest remained in the Ardennes, and this was an important reason why the transition was easier there. Increasingly effective and rational use of this property generated much income for both private individuals and communities (Tallier, 2004, Servais, 2013). This income helped keep down local taxes. Local landowners retained exclusive access to these commons, so local family transmission practices paid close attention to the strict physical division of the inherited property (Servais, 2003; Servais, 2005).

IV. Conclusion

On the basis of these three case studies, three conditions emerge as necessary, if not intrinsically sufficient, conditions for specialisation. First was a demand for specialized products, from a dense, growing and increasingly urbanized population. Second was the presence of physical markets, essential because they brought together producers, intermediaries and consumers to redistribute the specialized products, and also allow the producers to acquire items which they no longer produced themselves. Third were the means of communication and the transport logistics that made these markets effective.

However, these necessary conditions require a catalyst to develop. The catalyst could be exogenous: in the sixteenth and nineteenth centuries it was a political one. But it can also be endogenous, the simple result of evolving demand, a specific need or the overall conditions of production. Naturally it could also come from a complex mix of endogenous and exogenous factors. At the farm level, specialisation had a multiple impact, and required human and financial resources to sustain the process of transition and transformation. Between the eighteenth and the twentieth centuries, we can observe:

– a narrower range of produce, but not a general turning to a single crop;

– increasingly professionalized operations, trajectories and personnel, aided particularly by ever-increasing mechanization;

– a corollary increase in both in landed capital and working assets, but also a noticeably declining return on investment, except in the Ardennes region;

– the consolidation and reinforcement of local farming patterns : owner-farmers and smallholders in the Herve region and in the Ardennes, tenants and large farmers in the Hesbaye area;

– greater emphasis on the role of the farm within particular and diverse agricultural, economic, social and family systems;

– a generally rising standard of living, but among a population constantly reduced in size by migration out of the area.

Bibliography

BIHOT, Charles (1913), *Le Pays de Herve. Étude de géographie humaine,* Anvers, Van Hille-De Backer J., 170 p.

BRULARD, Théo (1962), *La Hesbaye. Étude géographique d'économie rurale,* Louvain, Librairie universitaire Uystpruyst, 306p.

DELATTE, Ivan (1945), *Les Classes rurales dans la principauté de Liége au XVIIIe siècle,* Paris, E. Droz – Liège, Faculté de philosophie et lettres (impr. de G. Thone), 337 p.

HOYOIS, Giovanni (1949-1953), *L'Ardenne et l'Ardennais. L'évolution économique d'une région,* Bruxelles-Paris-Gembloux, Éditions universitaires, 983 p.

MENDELS, Franklin (1972), 'Proto-Industrialization: The First Stage of the Industrialization Process', *Journal of Economic History,* p. 241-261.

NEVEN, Muriel & ORIS, Michel (2004), 'Les Industries rurales dans l'Est de la Belgique au XIXe siècle. Une histoire d'ambiguïté', in Jean Marie YANTE (ed.), *À l'approche d'une autre ruralité. Campagnes et travail non agricole du bas Moyen Âge à 1914,* Bruxelles, Archives et bibliothèques de Belgique, p. 91-124.

NEVEN, Muriel (2003), *Dynamique individuelle et reproduction familiale : le Pays de Herve dans la seconde moitié du XIXe siècle,* Genève, Droz, 530 p.

RUWET, Joseph (1942), *L'Agriculture et les classes rurales au Pays de Herve,* Liège-Paris, Bibliothèque de la Faculté de Philosophie et Lettres, 336 p.

SERVAIS, Paul (1982a), 'L'Évolution des structures agraires du duché de Limbourg du XVIIe au XIXe siècle', *Annales Économie, Sociétés, Civilisations*, p. 303-319.

SERVAIS, Paul (1982b), 'Structures agraires et industries rurales : le cas de l'Entre-Vesdre et Meuse aux XVIIIe et XIXe siècles', *Revue belge d'histoire contemporaine,* p. 179-206.

SERVAIS, Paul (1982c), *La Rente constituée dans le Ban de Herve au XVIIIe siècle,* Bruxelles, Crédit communal de Belgique, 390 p.

SERVAIS, Paul (1994), 'Marchés immobiliers herbagers et céréaliers au XVIIIe siècle: le cas de la périphérie liégeoise', in Michel DORBAN & Paul SERVAIS (eds), *Les Mouvements longs des marchés immobiliers ruraux et urbains (XVIe-XXe siècles),* Louvain-la-Neuve, Academia, p. 109-124.

SERVAIS, Paul (2003), 'Transmission patrimoniale dans l'est de la Belgique au XIXe siècle : Sart-lez-Spa et Clermont sur Berwinne', in Laurent HONNORÉ, Véronique FILLIEUX & Paul SERVAIS (eds), *Angles d'approches. Histoire économique et sociale de l'espace wallon et de ses marges (XVe-XXe siècle),* Louvain-la-Neuve, p. 125-135.

SERVAIS, Paul (2005), 'Transmissions patrimoniales en période de transitions industrielles : les campagnes liégeoises au XIXe siècle', in Michel BERTRAND (ed.), *Pouvoirs de la famille, Familles de pouvoir.* Actes du colloque international de Toulouse (5-7 octobre 2000), Toulouse, Université de Toulouse-le-Mirail, p. 805-816.

SERVAIS, Paul (2013), 'Common Lands and Agricultural Development in the foothills of the Ardennes 1750-1900', in Gérard BÉAUR, Phillipp SCHOFIELD, Jean-Michel CHEVET, Maria-Teresa PEREZ-PICAZO (eds), *Property Rights, Land Markets and Economic Growth in the European Countryside (Thirtheenth-Twentieth Centuries)*, Turnhout, Brepols, Rural History in Europe, 1, p. 139-156.

STATISTIQUE DE LA BELGIQUE, *Agriculture. Recensement général de 1895. Partie analytique (1899),* Bruxelles, Ministère de l'Agriculture, 560 p.

TALLIER, Pierre-Alain (2004), *Forêts et propriétaires forestiers en Belgique de la fin du XVIIIe siècle à 1914 : Histoire de l'évolution de la superficie forestière, des peuplements, des techniques sylvicoles et des débouchés offerts aux produits ligneux,* Bruxelles, Hayez, 764 p.

THOMASSIN, Louis-Ferdinand (1879), *Mémoire statistique du Département de l'Ourthe,* Liège, 487 p.

VLIEBERGH, Émile & ULENS, Robert (1907), *La Population agricole de la Hesbaye au XIXe siècle,* Contribution à l'étude de l'histoire économique et sociale, Bruxelles, 550 p..

VLIEBERGH, Émile & ULENS, Robert (1911), *L'Ardenne. Population agricole au XIXe siècle, Contribution à l'Histoire économique et sociale*, Bruxelles, Hayez, 428 p.

3. An early form of specialised agriculture in Western France

Martine Cocaud

I. Introduction

What do we mean by specialized agriculture in twentieth-century Western France? Corentin Canevet, in his study *Le modèle breton* (Canevet, 1992), describes Breton agriculture as being integrated into the market through intermediary consumption and the production of agricultural goods intended for sale. This integration occurred on medium sized farms, using mostly family labour, and was part of highly specialized networks of milk, pork, and vegetable producers. This model came into being in the 1950s and 1960s. The success – at least in terms of production figures – of this model has erased the memory of the previous one. The classic image is of traditional agriculture based on mixed farming, practised by individual families, principally for their own consumption. However, recent research has shown that the rationale behind the system of mixed farming and livestock raising was not archaic. On the contrary this system, whose results admittedly remained rather limited, made it possible to make the most of natural conditions and the West's outlying geographic position. Case studies, based for the most part on farm records from the eighteenth and nineteenth centuries, confirm this thesis. Farmers in the West wanted to live off the land, but they also farmed for profit, and tried their best to find a niche in the marketplace: this double objective required them to reconcile their production of grain and livestock. Drawing upon previously published work as well as upon the analysis of records from several farms in the department of the Ille-et-Vilaine, this study will show that, in spite of its reputation for resisting change, traditional agriculture possessed certain characteristics that one finds in specialized agriculture, such as strategies which differed according to farm location and size, and participation in markets. This article also reflects upon the impact of modernization on the specialisations of the past: while some were reinforced, others stagnated and did not take advantage of the market which was opening up at the time.

Agricultural specialisation and rural patterns of development, ed. by Annie Antoine, Turnhout, 2016 (*Rural History in Europe*, 12), p. 63-76.

BREPOLS PUBLISHERS DOI 10.1484/M.RURHE-EB.5.112261

II. A region of mixed farming and livestock

At first glance, 'the West' which is described in this paper – Brittany and a part of Maine – seems little specialized. It corresponded to a mixed-farming/livestock system dominated by heath and fallow land. It was a country of dairy cows and beef cattle, of rye and buckwheat. In spite of the considerable variations between the east and the west, and between coastal and inland regions, this area was largely homogeneous due, amongst other things, to the hedged farmland. The land was cultivated in the form of small plots, most of which did not belong to those farming them. The majority of farms were not owner-occupied; tenant farmers rented their land from rural or urban landowners In 1828, for the entire department of the Mayenne, there were 7,162 tenant farms, and 24,418 small enclosed plots or *bordages*, which implied a ratio of 1 tenant farm to every 3.5 small enclosed plots (Musset, 1917). The tenant farms were from 25 to 30 hectares in area, while half of the small enclosed plots measured from 10 to 15 hectares and the other from 6 to 12 hectares.

Throughout the West these tenant farms and *bordages* shared similar characteristics: the land which made them up was divided into three rings. Around the farm buildings was a first ring consisting of small plots (gardens, vegetable patches and orchards), always fenced in (Antoine, 1999a). A second comprised the ploughed lands – fields and plots – subject to crop rotation, regularly producing grain; and meadows. Beyond that stretched the poor-quality, little-cultivated land which became fallow, and then heath and rough grazing land. Outside the farms, the forests and heath were more or less left for collective use, depending on the region. These large uncultivated spaces were none theless useful in other ways.

II.1 Varied production on both a regional and a farm level

Small enclosed plots and tenant farms were laid out in the same way, the latter being a larger form of the former. All the farms put priority on grain production, using systematic crop rotation and making good use of fallow, but they also gave an important place to raising cattle or horses; the most common kind of livestock was dairy cows. Crop rotations began with buckwheat, followed by a summer grain (rye) and ending with a spring grain (oats) into which gorse was sown. The land was then left alone for several years. The herds grazed on these heaths and fallow; their droppings were mixed with the soil to make fertiliser, which was then spread on ploughed land. Livestock holdings in the West were in fact considerable: with a density of 27 heads per square kilometre, the department of Ille-et-Vilaine was at the French average. The poorest zones were where most livestock was found. Although dairy cows were most common, the least favoured areas practicing semi-extensive animal production also raised young steers, while the richest parts of the region often reared older oxen.

In looking at individual farms, it is clear that their organisation allowed both crop and livestock production. For example, on its sixteen hectares the Maison-Neuve farm grew buckwheat and rye (as well as wheat after 1850) plus some potatoes, which provided three fifths of the farmer's revenue (Cocaud, 1999a)[3] The other two fifths came from livestock products (animal sales and butter) and firewood sales. The farm had seven cows, four horses, four pigs, and sixteen sheep.

Figure 3.1. Map of location

Source. Cartography © A. Varet-Vitu-2016.

II.2. A logic of food production?

The combination of crop and livestock production made it possible to feed the local population. Grain had top priority because it was indispensable for this purpose. It was also the first to benefit from agricultural improvements, for in Ille-et-Vilaine the area of cultivated land went up by 25% between 1820 and 1860. However, all agricultural production was not intended for local consumption; much of it was sold through numerous well-established commercial channels. This was the case with butter: although most of the product was consumed locally, some was exported to Paris and to England as early as the seventeenth century. The prefect of the Côtes d'Armor put it this way: 'A considerable quantity of [butter] is exported abroad, besides the fact that no cheese is made in the department, but we can see from what has just been explained that the butter produced here (of excellent quality by the way), forms one of the most attractive branches of the rural economy and creates the wealth of the region'[1]. About a seventh or an eighth of locally-produced butter – all people at the time mention the difficulty of fixing the exact proportion – was eaten, either by its producers or by farm owners (butter was often among the products that a tenant farmer had to give to his landlord). The rest was sold.

The first stages of capitalism appeared early. In the West, sales took place partly at the market, where every week or two, women would bring butter already slightly aged and therefore salted. These markets served customers within a radius of 3 leagues, or even 5 to 6 leagues in the case of the largest. The buyers were individual consumers but also commission agents, often Norman wholesalers who were responsible for gathering assortments of butter and sending them to other regions. These wholesalers kneaded and mixed the butter together, sometimes melting and stocking butter from diverse origins: it then became what was known as 'strong' butter, that is to say strongly salted as it could contain up to half of its weight in salt[2]. Other butter was bought directly from farms by village grocers or by the employees of commission agents. The Norman butter trade with the large towns of the West was recorded as early as the sixteenth century, and in the eighteenth century the butter trade supplied not only the navy but also South America and Spain, for trade to the Caribbean. Butter was exported from the ports of Saint-Malo, Morlaix, Quimper, Nantes and Bordeaux.

Many other agricultural products were traded in much the same way, and there are many examples of widely marketed goods: cider, hemp, and grain, but also cattle and horses. This is then the proof that a traditional agricultural economy consisting of a balanced mix of diverse products fulfilled the food needs of the local population while at the same time furnishing considerable amounts to national and international markets.

III. Very diverse older specialisations

Closer examination of the different forms taken by livestock farming brings to light a wide variety of older specialisations.

1 Archives départementales des Côtes-d'Armor, 6 M 921: Information on bovine statistics ordered in 1819 by the Ministry of the Interior; answers from the department.

2 Archives départementales des Côtes-d'Armor, 7M 148, Fraud control. Letter from the minister of Agriculture to the prefect of Ille-et-Vilaine, August 1889.

III.1. A stated objective: the quest for profit

Until the end of the nineteenth century, Breton agriculture remained heavily constrained by its environment, with the result that rye-growing was restricted to acidic soil, wheat-growing to the coast, semi-intensive livestock raising to heathland and dairy farming to the Rennes basin; feeding the local population remained its main objective. But these limitations did not rule out the search for profit. The Society of Agriculture expressed it this way in 1757:

> Agriculture does not only consist in cultivating well, but in growing the things which will bring the most profit. The right choice in these matters depends on knowledge independent of agricultural practices. For example, those who own land far from the sea and rivers should not try to produce things whose volume and weight would require heavy transport costs. The most suitable use for their land would be for flax, hemp, and other crops which, by employing ample manpower, create added value, and this can compensate for the costs of freight. Most meadows should be situated near the sea, rivers, and convenient roads, because they make it possible to raise and maintain abundant livestock. Livestock gives its products every day to nourish mankind and also ensures much profit by supplying Butcher shops and Tanneries; and livestock can be transported over long distances without any other cost than its own feed.

It would highly desirable that in each district, intelligent people should examine the branches of agriculture which are or would be the most advantageous[3].

The need for profit therefore requires choices which bring into play not only the potential of the land and the know-how of farmers, but also the commercial environment. Analysis of farming accounts for holdings in different parts of the West shows that farmers, understood this point of view quite clearly and consequently put various strategies into practice. Grain, of course, accounted for the greatest share of all farm revenue, but the examples of specialisation presented here concern livestock, if only because stockraising practices bring out market strategies most clearly, and farming practice varies the most on this item. The examples show that the previously noted uniformity of the region concealed in fact a wide variety of local diversity.

III.2. The specialisation of tenant farms in Maine and in the Vitré region

In addition to its grain production, which brought in between 50 and 70 per cent of farm revenue, Maine was a region where cattle was raised up to the point between calf rearing and cattle fattening. Animals were normally born on the farms, lived there for a certain number of years, and were then sold to merchants from the Paris region or

---

3 *Corps d'observations de la Société d'agriculture, de Commerce & des Arts, établie par les États de Bretagne. Années 1757 & 1758* (Rennes, Vatar, 1760), p. XIV-XVI.

to grassland farmers in Normandy. Annie Antoine describes the livestock strategies which were deployed according to the different types of farms (Antoine, 1995):

> The dominant model prevailed on large farms (20 to 40 hectares) run by people invariably called 'tenant farmers' in the eighteenth century and more and more often 'farmers' in the first years of the nineteenth century. On these farms, the herd was made up mostly of males – bulls, steers and oxen – plus a few cows and their young, termed the 'pupils', that is, calves or heifers. The cows (farms had an average of four each) produced one calf per year; they were sold only when mature: ten, twelve even fifteen years of age. Among the newborn calves, all the males were kept and a female was kept from time to time to replace a 'worn-out' cow. Bulls were used once for reproduction (at the age of two), then castrated to become steers and then oxen, of which there were a dozen on a good-sized farm). Farmers from the Normandy grasslands would buy them once they had reached six to eight years, the 'age at which they fatten up the best': the farmer then had to feed the adult males for four or six years while waiting for merchants to seek them out. During this period, they were used for farm work and on the plough, but they were also frequently exchanged.

Antoine also describes a form of cattle speculation that followed from this:

> It consisted in delivering, especially to merchants from the Paris region) two- to three-year old heifers ready for their first calving. Certain farms seemed to specialise exclusively in this trade, while others put both six- to eight-year old oxen and two- to three-year old heifers on the market.

Two secondary options might follow from these two principal speculative models: farmers close enough to supply town markets on a daily basis might keep a dairy herd, selling off newborn calves as soon as they could so as to take advantage of increased lactation. Or they might produce large numbers of 'pupils' and sell them young. This model was found on the border of Maine and Brittany, where the Breton breed of cattle tended to predominate.

III.3. Specialisation in poor regions: extensive cattle raising near Redon

Heath and marshes predominated about Redon and in the eastern Morbihan; indeed heath on its own occupied up to half the surface in some communities. Part of the land was managed as commons and subjected to collective practices. The land was poor and communication difficult. All of the farms produced a mixed output and crops on the arable were rotated in strict rhythm, in this case buckwheat and rye followed by several years lying fallow. However the presence of a large amount of uncultivated land made it necessary to also raise cattle, sheep and horses. This was the region of Britany with the greatest number of heads of cattle per hectare in the eighteenth century, although the dairy cows were relatively few; there were 3 cows for

1 steers, as against the district of Vitré, which boasted 36 cows for one ox, Fougères, with 58:1, and Saint-Malo at 76:1). This district reared 70 per cent of the department's cattle. However, the technique used to raise them was completely different from that employed in Maine: animals drawn from small-scale local breeds roamed on the heaths, slept on hay all summer and were sold young. These young oxen were bought by breeders from the Loire-Atlantique or Vitré. The same was true of horses which were also sold very young, from the age of six months onward. Thus it was the natural moorland habitat which drove this region to specialise in calf production, but also the relative absence of communication and the poverty of farmers, which ruled out the practice of fattening their animals. Despite these constraints, livestock- raising was still well integrated into a market, certainly less lucrative than the one in Maine, but none the less durable (Cocaud, 1997 chapter 8).

III. 4. Dairy specialisation in the Rennes Basin

Dairy farming was practised throughout the West but it was in Ille-et-Vilaine that it obtained the best results. This was made possible by the presence of both heathland and meadows, though relatively badly maintained until the middle of the nineteenth century. The dairy cows, from local Breton or Norman breeds, were raised to produce milk, so their calves were always sold very young because profits on them were less than those coming from milk or butter sales:

> Small farmers hesitate to give up all the milk for the sake of the young calves… if the district supplies good butter and lies close to towns, it is in their best interest to sell the calf when at five weeks; more money can be got from the sale of the young calf and their milk and butter than can be made from raising the young animal[4].

Milk fed the family first of all, and the surplus was sold only in areas surrounding the towns. In the more remote countryside it was transformed into butter which fuelled the commercial speculation mentioned earlier.

III.5. Specialisations for profit?

A nuanced answer is called for here since it is obvious that all of these strategies did not have the same success. The farm accounts that have been analysed (Cocaud, 1999a) show that the ox and steer production characteristic of eastern Brittany procured good profits for large farms. These made out well by taking on both breeding and management: these farmers possessed cows whose milk they sold, otherwise they kept male calves which they fed and then sold at high prices when the animals had reached adulthood. Sizable sums passed through their hands and provided a

---

4 *Ibid.*

handsome income to tenants and landlords. However, when small farmers tried to play in the same league, the rules differed for want of resources: they would buy adult cattle which they would only be able to keep for a few months, either because they did not have enough feed reserves or because they could not tie up their capital for a long time. So they made meagre profits and were subject to economic slumps, when they could survive only because of their crops or with help from the landlord. Annie Antoine made the same observation based on farm accounts in Maine (Antoine, 1994) where farms seemed to be even better integrated into the cattle market. Although on all farms livestock profits seemed less unpredictable than those from grain, and thus provided a certain amount of stability, only tenant farms managed to earn substantial income. The latter were able to speculate on livestock, whereas the small enclosed farm plots remained very vulnerable. These people did indeed try to obtain maximum profit but could only approximate to specialisation by "copying" better-equipped farmers as well as their means would allow. This business secured some assets but did not bring about great results.

The same went for areas which were more distinctly oriented toward dairy farming. Statistical analysis of livestock makes the West appear to be a great dairy region with well-established commercial structures but this degree of specialisation seems less obvious on the level of individual farms (Mayaud, 1997). It was evident on the farms located on the outskirts of Rennes which made good money from the sale of butter and milk. However, other farms contented themselves with raising a few cows in order to consume their butter and milk. They sold the surplus and a few young calves but this business only made them a few pennies, although it was in their best interest to get these pennies on a regular basis. In the same way, extensive raising of small cattle on the heaths made very meagre profits. Through merchants, these farms participated in commercial circuits, but they did not necessarily enjoy the solid income which would have allowed them to improve their strategy. Are these scenarios a sign of the presence of a specialized agricultural model?

IV. The future of older specialisations in the nineteenth century

Although the system of cultivation in the West was not changed dramatically by the introduction or elimination of new crops in the nineteenth century, it was nevertheless modified. The area of meadows greatly increased, reaching two-thirds of grazing land in 1841. The decline in fallow and the spread of irrigated natural meadows and, to a lesser degree, the growth of artificial meadows where clover and rye- grass prevailed, all speeded up changes in crop rotation, though without totally upsetting the old system. These agricultural transformations worked to intensify livestock production.

The number of animals increased everywhere except on the coast, and the spatial distribution of different types of livestock management was changed (Cocaud, 1996).

IV.1. Certain livestock-raising regions became more specialized

The surface under meadows rose markedly on farms in the east of the department, reaching two-thirds of grazing land in 1841. As a result, some farmers in the neighbouring regions of Maine, who in the eighteenth century had kept some dairy cows and practised modest calf-rearing, opted decisively to join commercial beef circuits during the nineteenth century. They 'bought thin cattle, at the age of 15 to 18 months. The bulls were studs, castrated at two years – for a few years they had been kept as studs until three years old; as soon as they were castrated, they were brought into farm work. A large cattle trade existed in the region at the time. Farmers changed oxen ten, fifteen or twenty times a year... They bought slim cattle in Nantes, Rennes, Vitré and Maine. The trade may well have concerned 30,000 heads of cattle a year' (Bertin et Maupillé, 1846).

This change may be clearly seen in the farm accounts of the agriculturalist Marquis d'Argentré, who encouraged his farmers to adopt agricultural innovations from early on. The first attempts can be seen early as 1813 on the Étevinière farm[5]. In order to free himself from long-standing financial difficulties and stabilise his operations, the marquis ended the rent contract and had a sharecropping lease drawn up. He encouraged and financially helped his tenant to replace the cows, selling the oldest animals at low prices (47 francs on average) and buying high-value specimens (130 francs); he also speculated skilfully on the rest of the livestock in order to secure maximum profit. Each year on average, five animals were sold – one steer, one ox, one cow, one heifer, one calf – and two new animals were bought. Four to five calves were born a year, one was sold, another died, the rest were kept: one would become a heifer and be sold around one year of age, one or two males were kept to become steers; occasionally calves were bought to preserve a balance between the farm's males and females or to replace a deformed animal. In total, the Marquis' trade in young cattle made up 12 per cent of his income from livestock.

The most valuable operations concerned males: five steers and thirteen oxen were sold over eight years. The sale of the oxen provided 60 per cent of the farm's profits but income was irregular. Sometimes the farmer had little to show for his labour: thus, in 1817, he sold four oxen for 810 francs (202.5 francs each on average) and bought two more for 330 francs (for an average of 165 francs each) which he sold for

[5] Accounts of the d'Argentré family, Archives départementales d'Ille-et-Vilaine, non-classified collection.

a bit more the following year. However, most of the time the animals which were sold had been raised on the land for several years: the oxen were more than three years old and had been trained for ploughing, so their price was four times higher than that of the steers. In total, the profit from livestock rearing came to 3,167 francs over eight years[6], and the farm seems to have turned a profit in the end. Nevertheless, a year-by-year analysis of the accounts shows the instability of this income. In 1813 almost all of the livestock was sold, so the following year the farmer replenished it, which led to a deficit. In 1815 the profit was slim and it was non-existent in 1816 because nothing was sold. Sales did not really pick up until 1817; they ran out of steam in 1819 and results only seemed to be satisfactory from 1820 onward when profits began to surpass 450 francs a year. Only the large tenant farms that shared profits and losses between the farmer and the landlord could take a risk on this sort of specialisation; of course it was totally impossible for small enclosed plots.

IV.2. Other regions changed direction

This was the case with the Morbihan and in the east of the department of Ille-et-Vilaine. This region was poor, and there was semi-extensive raising of young animals on the heaths for market. Agricultural modernisation was implemented very slowly and in 1870 this still occupied 70 per cent of grazing land. Yet, in districts which managed to reduce the area under heath, most often for grain and feed production, the number of dairy cows increased noticeably. Only the most infertile zones persisted in extensive livestock production and made 'the greatest possible profit' from it[7]. Young oxen from eighteen months to two years of age were sold at fairs. They were bought by small farmers for ploughing, trained and then re-sold after a year to a larger estate which, in turn, kept them a year before selling them to grassland farmers from Vendée and Normandy for fattening.

So in this area, nineteenth-century agricultural improvements did not reinforce specialisation. To the contrary, traditional stock-raising was abandoned in favour of some dairy farming and especially for grain production. This is then a totally different configuration from that which Jacques Mulliez reported in his article 'Du blé, un mal nécessaire' (Mulliez, 1979) where the farmers of Normandy turned decisively toward livestock when they found themselves in a position to replace heathland with fields. In this part of the West, the natural infertility of the soil plus its geographical isolation made grain and livestock production the initial priority in agricultural modernisation. The hope of turning a profit took a back seat to the need to feed the population.

---

6 By comparison, the farm lease cost 400 francs per year in 1810.

7 Archives départementales du Morbihan, 6M, Enquête de 1906 (Survey of 1906).

IV.3. Failure in the dairy industry of Ille-et-Vilaine

Some farmers put more emphasis upon a specialisation already well underway, but failed to see it through. Between 1850 and 1890, the richest districts of Ille-et-Vilaine increased the area of their meadows and stock-raising, which grew at the same rhythm as that of French dairy farming as a whole. Moreover, in 1880, the number of Breton cows made up between 11 and 13 per cent of French livestock although Brittany only covered 6 per cent of cultivated area nationally. The pre-eminence of cows over other bovines tended to increase and the head of the Rennes agricultural show noted in 1866 that 'the farmer earns something from livestock and the sale of animals but butter is his clearest profit… thanks to butter we did not suffer from low grain prices as much as other regions, even though we met with great difficulty'[8].

Was Brittany then going to turn to specialisation, now made possible thanks to agricultural progress? To do this, it would have been necessary to greatly increase the number of cattle and the area of fields, and change the structure of milk collection from the collection of butter to that of cream and then of milk. This evolution came very late in the nineteenth century. Although this opportunity convinced a few agronomists to create a dairy school for girls (Cocaud, 1999b), it struck no chord with farmers. The regional increase in dairy livestock should rather be seen as the result of each individual farm pursuing higher profits. Several facts prove this point. Mixed farming and livestock-raising structures were not called into question, since the acreage sown with grain did not shrink until after 1929. Farmers also refused to sell milk and continued to make and sell farm butter. Changes would have had consequences for the methods of cultivation, the provincial economy and way of life: growing less grain; giving up the pigs that all farms raised thanks to whey, a by-product of butter; women losing the few centimes to be earned from the sale of butter; and the dismantling of existing commercial circuits. Although it remained a region of livestock-raising, the West turned its back on a more intense specialisation until 1950, when specialisation was forced on it, not by nature but by the imperatives of the new economy.

V. Conclusion

We can now define more specifically what is meant by specialisation in a region of hedged farmland (what is called bocage in France). In a traditional economy, a mixed-farming/livestock-raising system and specialized agriculture cannot be imposed straightaway, because whatever the strategy, farmers' need to eat and live

8 Ange de Léon, in charge of the Northeast Rennes agricultural show survey of 1866 (Archives départementales d'Ille-et-Vilaine).

made food-producing agriculture indispensable. Of course, the need for fertilizer also implies the obligation to raise livestock, the famous 'necessary evil' of nineteenth-century agronomists. However as several historians have shown, reality was more complicated than that[9]. Farmers did not indiscriminately raise livestock, and the animals themselves were much more than manure suppliers. Choices were made – extensive or intensive livestocking, for meat or milk – in a rational manner and taking heed of limitations. One of the first forms of agricultural specialisation can be seen in these choices, which were linked to natural constraints. The management of livestock is what demonstrates the role of specialisation in the traditional mixed-farming/livestock-raising system of the West. This activity essentially depended on the market and allowed farmers to procure some cash for themselves.

However, certain forms of livestock raising were more endured than chosen, a sort of forced choice, as contradictory as these two terms may seem. The interest of these forced specialisations was sometimes called into question with the progress of agriculture. Whereas the rearing of trained oxen in Mayenne, a profitable activity from the eighteenth century onwards, continued, with constant improvement, until the late nineteenth century, extensive raising of small oxen in the Redon area disappeared as soon as agricultural progress permitted. Might it be said that this poor man's version of livestock-raising – a completely rational activity, but one with little economic point – could be called a specialisation? Currently, specialisation is associated with high productivity and the will to make a business endure by perfecting it. Would the raising of small oxen have then been a makeshift solution for a poor region? Could this have been specialisation by default?

We may also wonder whether specialisation ever really interested Breton farmers before 1950. There could be doubts on the subject. Perhaps they never asked themselves the question because surveys and reports never presented agricultural progress in terms of specialisation. Jacques Mulliez confirmed this point of view:

> New agriculture from 1750 onwards tended in one way or another to combat the environment. What the majority of agronomists wanted for grain-producing regions was that they remain that way, but that they also become livestock regions, and inversely that livestock regions would become grain regions as well. Ideally each landlord, whatever the situation of his farm, would be able to use the rules drawn up by the science of agronomy in order to produce what he believed to be the most profitable without being a prisoner of his environment. Traditional agriculture and livestock raising were created from adaptation to the environment. (Mulliez, 1984).

[9] The first questioning of cattle as a 'necessary evil' was without a doubt Mulliez' article (1979) whose demonstration was based on reports, memoirs and surveys. Antoine (1994) showed once again from the analysis of farm acounts, the benefits of livestock on tenant farms in the West.

Yet in the later nineteenth century, a few individual attempts occurred under the pretext of modernisation and profitability in the face of the grain crisis. In Ille-et-Vilaine, in the 1870s, the agronomist Comte de la Riboisière set up around twenty model farms in order to supply a modern dairy. He gave each of his farmers a new herd made up of dairy breeds. However apart from these, he did not convince the district's farmers to participate in milk collection and abandon butter-making on their farms. It was not until after 1950 that the West would make the choice of specialized agriculture and turn toward dairy farming. This refusal of specialisation was found in other sectors: the production of cider apples for example. Why this rejection? The reasons for it are complex but it seems that at the beginning of the twentieth century, dairy specialisation had not found legitimacy from the perspective of a new rationality and a larger potential profit.

Precocious dairy specialisation in the West, which generated intense commercial activity, slowed down when faced with the possibility that it might have to change. As a result, the Charente region, despite a less firmly established dairy tradition, turned much more easily towards the butter industry and deprived Breton commerce of its leading position in this trade. In view of this, Henri Sée supposed that the refusal to change had originated not with farmers but with wholesale merchants who did not want to threaten traditional salted butter commercial circuits (Sée, 1921-1923). This hypothesis later drew criticism[10] but it is interesting because it suggests a new way in which various key players interacted in the economy when faced with agricultural innovation. Whereas in the present day the place of middlemen in the agri-food economy is always taken into account, they come less spontaneously to mind in in discussions of earlier periods – but they were there. Could their ability to influence agricultural trends not be considered proof of the presence of specialized agriculture?

Bibliography

ANTOINE, Annie (1994), *Fiefs et villages du Bas-Maine au XVIIIe siècle. Étude de la seigneurie et de la vie rurale dans une province de l'Ouest au dernier siècle de l'Ancien Régime,* Mayenne, Éditions régionales de l'Ouest, 536 p.

ANTOINE, Annie (1995), 'Les Bovins du Bas-Maine, un modèle économique pour les campagnes de l'Ouest ? (1720-1820)', *Histoire et Sociétés Rurales,* 4, p. 105-136.

[10] Guillou (1988) supposes that women wanted to keep the production and sale of butter at the market that got them a bit of cash. Henry (1966) agrees in part with this analysis and links the refusal of industrial dairies to the very large number of cream separators found on Breton farms. Proof that a technical object which did not threaten social organisation could quickly be adopted. The fact that the abandon of farm butter implied difficulties for raising the pigs that were found on all farms could also in part explain the refusal. Indeed, pigs were fed with whey mixed with potatoes.

ANTOINE, Annie (1999a), 'Systèmes agraires de la France de l'Ouest : une rationalité méconnue ?', *Histoire, Économie et Société*, 1, p. 107-132.

ANTOINE, Annie (1999b), 'La Sélection des bovins de l'Ouest au début du XIXe siècle : évolution des pratiques et des représentations', *Annales de Bretagne et des Pays de l'Ouest,* 1, p. 63-86.

BERTIN, Amédée, MAUPILLÉ, Léon (1846), *Notice historique et statistique sur la baronnie, la ville et l'arrondissement de Fougères*, Rennes, A. Marteville et Lefas.

CANEVET, Corentin (1994), *Le Modèle agricole breton, histoire et géographie d'une révolution agro-alimentaire,* Rennes, Presses Universitaires de Rennes, 518 p.

COCAUD, Martine (1999a), 'Une Approche de la modernisation agricole de l'Ouest au XIXe siècle : des statistiques aux comptabilités d'exploitation', *Ruralia,* 5, p. 7-31.

COCAUD, Martine (1999b), 'L'Avenir de Perrette. Les premiers établissements féminins d'agriculture : les écoles pratiques de laiterie', *Annales de Bretagne et des Pays de l'Ouest,* 106, 1, p. 121-135.

GUILLOU, Anne (1988), 'L'Argent du beurre', *Ar-Men,* 16, p. 18-25.

HENRY, Jean-Baptiste (1966), 'L'Industrie laitière en Bretagne et les révolutions techniques du XIXe siècle', *Annales de Bretagne et des Pays de l'Ouest,* 73, 2, p. 255-281

MAYAUD, Jean-Luc (1997), 'L'Élevage bovin : d'un mal nécessaire à la spécialisation', in Monique PAILLAT (ed.), *Le Mangeur et l'animal*, Paris, éd. Autrement, p. 11-33.

MULLIEZ, Jacques (1979), 'Du Blé, mal nécessaire. Réflexion sur les progrès de l'agriculture 1750-1850', *Revue d'Histoire Moderne et Contemporaine,* 26, p. 3-47.

MULLIEZ, Jacques (1984), 'Pratiques populaires et sciences bourgeoise : l'élevage des gros bestiaux en France de 1750 à 1850', *in L'Élevage et la vie pastorale dans les montagnes de l'Europe au Moyen Âge et à l'époque moderne,* Clermont-Ferrand, Institut d'Études du Massif Central, p. 289-304.

MUSSET, René (1917), *Le Bas-Maine. Étude géographique,* Paris, Armand Colin, 496 p. (Laval, Librairie Cantin, 1978).

SÉE, Henri, (1935), 'L'Industrie et le commerce de la Bretagne dans la première moitié du XVIIIe siècle d'après le mémoire de l'Intendant des Gallois de La Tour', *Annales de Bretagne*, 35, 3, p. 433-455.

4. Agricultural specialisations and the sea, seventeenth to nineteenth centuries

Gérard Le Bouëdec

I. Introduction

Agricultural production was the main cargo of seaborne trade. With the expansion of colonial and intra-European trade in the first wave of 'globalisation' during the seventeenth century, international demand played a part in transforming many agricultural activities of the coastal hinterland. Clearly the attraction of markets in northern Europe, dominated by the Dutch, the English, and later the Scandinavians, helped expand Atlantic and Mediterranean vineyards, reducing the extent of of cereal crops, or striking a new balance between crops planted on the same ground. The drive to increase wine production did not necessarily bring about new planting of good quality vines. Although wine-growing took on a speculative character, it was impossible to avoid the need to grow grain until at least the end of the eighteenth century. In the Vannes region, close to the coastal ports of south-western Brittany, grain was grown almost exclusively and surpluses were frequent. The ups and downs of the agricultural economy in south-western France and the Iberian Peninsula left some openings for the coastal agriculture of South Brittany. But during the eighteenth century, the new opportunities opened up in interregional markets brought on very few changes in the way production was organised. In the nineteenth century, the grain market lost its speculative character and the Gulf of Morbihan stopped playing the role of wheat granary.

Around 1850, the focus of innovation moved from the mainland to the foreshore. Shoreline populations had always made much use of the foreshore, and farmer-fishermen dredged the shallow waters to the point of exhausting the oyster stocks. Now they were forced to innovate by becoming sea-farmers. Oyster-farming was the first form of aquaculture singled out for development and was put under the authority of the Naval administration. However the vocabulary used to describe the zone of cultivation and the organisation of concessions betrayed its agricultural origins. It

Agricultural specialisation and rural patterns of development, ed. by Annie Antoine, Turnhout, 2016 (*Rural History in Europe*, 12), p. 77-98.

BREPOLS PUBLISHERS DOI 10.1484/M.RURHE-EB.5.112262

was no coincidence that in 1875 the pioneers in this field chose a regional agricultural show as the forum in which to display their first successful results.

In fact, the study of coastal agriculture has too often been neglected. Although coastal zones were not necessarily zones of specialised agriculture, in general, the sea took second place to the land there. Yet 'sea manure' helped agricultural production flourish. Indeed, soils enriched in this way were raised to such a high quality that, in the second half of the nineteenth century, a coastal 'golden belt' gradually developed, specialising in early vegetables, specially targeted at urban markets.

Figure 4.1. Map of location

Source. Cartography © A. Varet-Vitu-2016.

II. The expansion of Atlantic and Mediterranean vineyards in the seventeenth and eighteenth centuries

When wine growing becomes more than just a luxury and is done for commercial profit, it needs to be located in places where the product can be easily marketed [...] Thus, it is not surprising [...] to find famous vineyards and seaports so often located in the same place and for so long a time (Dion, 1959: 49).

Gilbert Buti, in his study of the Provençal coasting trade, quotes Joseph Fauchet, prefect of the Var at the beginning of the nineteenth century: 'The best place to sell [wine] is near [the sea]' (Buti, 2007: 10; Fauchet, Year X [1801-1802]: 21). Nor did the wine have to be of high quality for it to find a niche in seaborne trade networks.

II.1 The vineyards of Nantes and Aunis-Poitou: less grain and more wine for brandy distilling

At the end of the seventeenth century, intendant Béchameil de Nointel observed that 'over the last several years, many vines have been planted in Brittany...' (Dion, 1959: 456). Roger Dion pointed out that the authorities intervened to halt the excessive spread of these vineyards, as they were depriving Anjou vines of their market share. However, the main reason why officials took action was because the area devoted to growing grain was shrinking, driving grain production down and prices up.

The restrictions on planting vines in Brittany, particularly throughout the Nantes region, were put in place on the King's orders. The reason was that the inhabitants of those regions, seeing that the wine and brandy trade was profitable, nearly stopped sowing wheat altogether, with two unfortunate effects: one, that grain in those parts became very dear and often insufficient, the other that wines and brandy from Anjou [...] were no longer shipped out by sea (Dion, 1959: 456).

This decline of grain crops, and their replacement by wine growing, were brought about by the Dutch. Intendant Béchameil de Nointel noted that wines so produced were not of high quality, and that they were grown to be distilled into brandy. The Dutch encouraged the planting of gros plant vines, poor in quality, but high in yield; they produced 50 to 60 hectolitres a hectare as compared to 35 to 40 for the better-quality Muscadet. Gros plant had the advantage of being cheap and abundant.

The salt marsh regions all focused on distillation, particularly the islands of Oléron, Noirmoutier and Ré. According to historian Michel Albert Luc, vines took up 4,520 hectares on the Ile-de-Ré, or about two-thirds (61.93 per cent) of its surface, being particularly dominant in the Baronnie area, where they occupied 83.6 per cent of the land (Luc, 2005: 150). The cartographer Claude Masse, in a 1712 treatise on the island's toponymy, noted that in 'some years more than 55,000 barrels of wine are laid up on the Ile de Ré, almost all thin white wine, with a nasty smell and bad taste, like the sart or wrack which are applied to them as manure' (Lachiver, 1988: 259). The Dutch set up distilleries in the Nantes countryside and bought directly from producers, thereby short-circuiting local trade networks (Collins, 1984: 275). In Oléron, the earliest distillery dated from 1604; in the Ile de Ré the first was built in 1708 by Baron Le Berton de Bonnemie (Guillemet, 2000: 58). Even though the local wines had little merit, the Nantes and Aunis-Poitou brandies produced from them were particularly prized in Bremen (Vos, 1995: 184). Pinczon du Sel des Monts, in a contemporary description, observed that 'of all the brandies made in France, those from Nantes have the finest taste, smoothness and strength and because they have the best 'beading'[1].

French ship-owners bear their cargoes to foreign destinations, especially the West Indies, Canada, Guyenne, the African coast and Northern Europe, but it is also reckoned that foreign ships, English, Scottish, Flemish, Hamburg, and particularly Dutch, that come for cargoes, take seven to eight thousand barrels every year' (Pinczon du Sel des Monts, 1756: 48). The increase in outlets in Northern Europe, the Baltic and in thecolonial trade made winegrowing a bigger concern. Thus, on the Ile-de-Ré, the entire population invested in vines, not just ploughmen-wine growers but also people in trade, and even seafarers and craftsmen. Agricultural and salt production thus helped integrate the island's economy into the European economy. Despite the importance of the home market, which took on cargoes and victuals for colonial shipping, the production of the Nantes and Aunis vineyards continued to depend on the market in Northern Europe. However, from the 1760s, the shipping role of the Austrian Netherlands and the United Provinces was gradually taken over by the Hanseatic ports (Lübeck), and then by those of Scandinavian countries, Sweden (Stockholm and Göteborg) and Denmark (Copenhagen). This restructuring of markets and trade caused a good deal of volatility and finally a fall in brandy exports from the 1770s onward.

II.2. Mediterranean coastal vineyards: from secret cultivation to speculative cultivation

All wines are not high quality and some cannot travel by sea, as in the case of Languedoc wines such as Clairette, Muscat, Picardan, Saint-Georges and Saint-

[1] POURCHASSE (2003: 342) defines beading (*l'épreuve de chapelet*) as 'a small white froth that forms a ring around the surface of the brandy when it is poured into a pitcher and makes a circle as it goes down – the hallmark of good brandy'.

Christol, as well as the Rhône wines of Tavel and Roquemaure; nevertheless wines from the Provençal terroirs of Sanary, Bandol, Ceyrest, and Cassis were still all considered quality products (Buti, 2007: 12). The development of vineyards should be seen in the context of the agricultural landscape. According to René Baehrel 'in many places, especially those on the eastern side of Lower Provence, wheat was no longer the main produce' (Baerhel, 1961: 91). Most important, conditions of production changed. Traditionally, vines shared the same soil with cereals and fruit trees such as the olive; they were set out in linear fashion, separated by ouillères, which were three-metre wide spaces of tilled land bearing grain, vegetables, or fruit. Under these conditions wine was treated as a supplement to other crops. Once it developed into a commercial product it became a crop in its own right, taking over all of the land allotted to it in 'solid vines'; near Toulon and in the Saint-Tropez peninsula, it took up 80 per cent of the land. Gilbert Buti notes particularly how it increased the practice of using seaweed as fertilizer, placed at the base of the vines, although less often than in the Atlantic coastal vineyards (Buti, 2007: 24). The land surveys for Ramatuelle, Gassin, Cogolin and Saint-Tropez give evidence of the interest seafarers took in wine-growing. In Saint-Tropez, seafarers controlled 40 per cent of the wine-producing land; and even though captains and ship-owners amounted to only 7 per cent of the local population, they possessed a quarter of the land.

This passion for wine-producing land was just as evident in Languedoc, where it was said around 1770 that 'in this very year and three or four years afterwards, there were vineyard owners in Marseillan, in the diocese of Agde, who earned as much from a single year's harvest of fruit as the plots of land they used were worth; quick fortunes are made in the wine-growing parts of the province and among the traders of Sète… In this way almost all the properties bordering on the sea were soon planted with vines, and young traders are sent from the port of Sète to the northern provinces of France and to foreign countries to advertise the wines and brandies of Languedoc' (Dermigny, 1955: 33).

The basic explanation of this growth was demand from nearby urban markets, particularly Marseille, where the population was growing rapidly in the second half of the eighteenth century, and also the naval port of Toulon as well as the seaboard cities of Genoa and Livorno further east. But a wider range of agricultural products also fitted into the local economy of agriculture and the coastal trade. As Gilbert Buti aptly puts it – borrowing a quote from Fabry, Director of the Naval Dockyard in Toulon – the 'path of the sea' was 'the foster father' of France's southernmost provinces (Buti, 2000: 297-320). True, as early as 1698 a memorandum from Intendant Lebret described 'the plains that reign from Aix to Marseille and to the river Var' as being 'covered with orange and lemon trees, olive trees, fig trees, pomegranates, and

[...] filled with the finest vineyards in the world.' However this entire 'fruit basket' region (Buti, 2007 and forthcoming) moved beyond tsimple production of food for consumption when it began to grow lemons, in response to specific requests from the Navy for medicines to counter scurvy, and when proto-industrial leather producers in Grasse were called on to help make scented leather for clothing. In this way a burgeoning perfume industry (Buti, 2007: 7) was reinforced by the sowing of jasmine, lemon, orange, and citron trees. This trade was left to the micro-entrepreneurs of the coasting trade; Marseilles shippers hardly invested in it at all. Providing wine, fruit and sweet chestnuts became the task of the smaller units in the port network, their ship-masters specialising in these cargoes or, like the sailor-winegrowers or winegrower-sailors, moving their own produce along the coast.

Regional dynamics alone do not account for the growth of the Mediterranean vineyard. Wine and brandy were part of inter-European trade, with ramifications in European colonial trade as well. Thus, the development of Languedoc and Toulon vineyards owed much to Dutch demand, but in the late eighteenth century, more precisely around 1770, vineyard expansion was linked to the reconfiguration of trade in northern Europe as a whole. Pierrick Pourchasse (Pourchasse, 2003: 103) has demonstrated the parallel between falling exports from Bordeaux to northern Europe and the Baltic, and an increasing outflow of wine from Languedoc in the form of Picardan, used to produce muscatel. According to the Swedish customs-records for the Sund, Langedoc became the second centre for French exports to the Baltic. The Swedes were the principal customers for Languedoc products, Pourchasse finds that they accounted for 30 per cent of shipments to the Baltic between 1783 and 1788. The port of Stettin was their second biggest destination, on board Swedish and Danish vessels. Languedoc wines from the Mediterranean port of Sète were cheaper than Bordeaux wines and fitted perfectly into Scandinavian trade with southern Europe, as they could make up return cargoes for Northern ships coming back from the ports of Marseille, Toulon, Livorno and Genoa. Thus the rise of demand for the produce of Languedoc wines at the expense of those from Bordeaux had less to do with the quality of the product than it had with reconfigured trading patterns. The substitution of the Swedes for the Dutch in the trade with southern Europe shifted its centre of gravity towards the western Mediterranean. Swedes sailing to Mediterranean ports, like other Scandinavian shippers, preferred to load wine, and even their salt, rather than make a second stop in Bordeaux (Gaussent, 1987: 89).

III. From grain growing to oyster-farming in the Gulf of Morbihan: The agro-maritime grain system in Vannes in the eighteenth century

T.J.A. Le Goff's study of the Vannes region in southern Brittany defines perfectly the local economic model:

> Essentially, the Vannetais in the eighteenth century provides an example of the 'one-cro' export economy characteristic of many developing countries today, and, again, as in many of these countries, it tended to benefit the more prosperous sort of merchants, those who, as landlords, prosperous peasants or rentiers, held large stocks of grain and, more generally, the mercantile town over the agricultural countryside (Le Goff, 1981:14; 1989: 30)

Local micro-capitalism was thus created around Vannes, which regularly benefited from a surplus of winter wheat cultivated on the coast, and of rye, from parishes further away from the littoral. In 1698, the Intendant, Béchameil de Nointel, reported that in an average year 9,000 tonneaux (155,000 hl.) of rye and 6,000 tonneaux (103,000 hl.) of wheat were exported from the three ports of Vannes, Auray, and Hennebont, most of it, some 4,000 to 5,000 tonneaux, coming through the port of Vannes. But Le Goff considers that this figure was often greatly surpassed, and that Vannes was the second most important Breton port for grain exports after Nantes. Most prominent in the trade were semi-specialised merchants such as Dubodan, La Rive, La Motte, Delourme, and Desruisseaux, but the entire landlord class too were involved in this speculative agricultural economy.

The produce found markets in the Bay of Biscay, sometimes Portugal, sometimes Spain, occasionally in the western Mediterranean, but chiefly in Bordeaux; in 1748-51, the biggest local merchant in the market, Advisse Desruisseaux directed 64 per cent of his gross turnover in dealings with grain firms to Bordeaux, though much of this was re-exported clandestinely to Spain and Portugal (Le Goff, 1973: 99; Barthélémy, 2002: 40).

Table 4.1 demonstrates the risks on this market, which was still speculative. As can be seen in this table, the margins were smaller in the 1730s, and specifically after 1770. The best times were between 1737 and 1746, and from 1753 to 1768. To supply the Bordeaux and northern Spanish markets, Vannes could count on an exportable surplus of 4,000 to 5,000 tonneaux, which implied a sales turnover of 800,000 to 1,000,000 livres a year, assuming that some growth took place. This was considerable, given that shipbuilding in that this area, a major centre for this activity, earned a gross income of only 20 to 30,000 livres annually.

Table 4.1. The growth of profit margins on wheat

	Vannes to Bordeaux	*Vannes to Spain*	*Bordeaux to Spain*
1728-1743	-0.9	1.4	2.4
1733-1748	1.9	0.5	-1.9
1738-1753	0.0	0.9	1.1
1743-1758	-2.3	-2.5	0.5
1748-1763	0.0	0.4	0.4
1753-1768	1.8	1.1	-0.7
1758-1773	-0.5	0.1	0.6
1763-1778	-0.4	-4.8	-3.6
1768-1783	-0.3	0.3	1.0
1773-1788	-0.5	1.7	1.8

Source. Le Goff (1989: 298).

There are some signs of increasing production in the eighteenth century, during the 1720s and the 1780s. Harvest estimates drawn up by the local representatives of the provincial intendant suggest that rye production rose by about 0.4 per cent a year and wheat production by 0.5 per cent between 1734 and 1788. Le Goff mentions some modest improvements – a few hundred additional hectares cleared during that time, some increase in output, and a change in crop rotations (Le Goff, 1981:293-295; 1989: 262-264), but maintains that this rise was more a return to levels reached during the previous century, aided by a long run of good weather. There were, therefore, no major changes in the production system despite the general increase in demand suggested by rising prices of around 1.2 per cent per annum for winter wheat and 1.5 per cent a year for rye. It was rather imbalances in supply, demand and price between this and other regions that provided openings for the micro-capitalists of Vannes. The merchant tried to pick up as much grain as he could from farmers, who needed to sell very soon after the harvest, at a relatively low price, so as to get cash to pay their rents and royal taxes. But the bulk of exported grain came from contracts with ecclesiastical and lay landowners, most often with their agents like a certain Le Corre, who on behalf of the Comte de Châteaugiron contracted in 1777 to supply 125 tonneaux, worth a total of 3,053 livres.

However, this cereal-producing mono-activity made the Vannes region vulnerable, as the intendant's representative pointed out in 1758: 'When there is a lack of outlets, those who are best off find themselves in a sort of indigence […] because even though harvest fluctuations ought to affect the price of this foodstuff, such circumstances affect it less than a shift in exports' (Le Goff, 1981: 326; 1989 297). There are, nevertheless, some clues to the social and economic impact of rising exports from the 1750s to the 1770s. The price of grain from Vannes went up by 30 per cent between 1764 and 1775. This 'boom', as Le Goff terms it, was more profitable for tenants than landowners, as tenant income increased faster than rents. Cultivable land was in short supply, which, in this region of 'repossessable' land (domaine congéable), showed up in increasing numbers of tenant evictions from 1767 to 1777. Most land was held by tenants under domaine congéable[2] which no doubt made it more difficult for bourgeois to acquire rural property; despite some bourgeois incursions into the property market, the bulk of property remained in the hands of the nobility, who were legally the proprietors of the land under domaine congéable. But in the end, rents and taxes caught up with the rise of prices around 1772, and grain from Vannes could no longer compete on the Bordeaux market. The Vannes region depended heavily on its land, which allowed an intermittent accumulation of capital, but neither merchants nor landlords could get through many years without exports. The career of the merchant Advisse Desruisseaux points up the fragility of this system. He was a commission trader specialising in trade with Bordeaux. He built up a network of suppliers among the landholding clergy, nobility and official class who speculated by putting their stocks of grain on the market after lesser folk had sold their produce. As Desruisseaux himself did not have a great deal of capital, he was forced to request advances from his local suppliers and his Bordeaux correspondents, which he could not reimburse until he had been paid for delivery. All the different parties were linked together through the bills of exchange which Desruisseaux's purchasers in Bordeaux and elsewhere allowed him to write against their accounts with Paris bankers and sell to these local financiers. These financiers had large sums of cash at their disposal, which was what suppliers wanted for their grain. Desruisseaux used the services of, among others, Nicolas Viel, municipal treasurer and collector in the 1740s, his own brother-in-law J-N. Provost, director and receiver for the royal Domain in Auray, his wife's cousin Penquer, sometime Director General and controller of the Domain in the Vannes diocese, subsequently receiver-general of Breton duties in the 1730s, and the two Guérins (one his brother-in-law and the other his nephew) warehouse-keepers of the royal tobacco monopoly in Vannes and Auray.

---

[2] *Domaine congéable* is a particular form of tenure, very common in western Brittany. The lease distinguished between the landlord (*seigneur foncier*) and the tenant, called variously the *domanier, tenuyer, colon* or *édificier*. The tenant owned the buildings, ditches, banks, planted fruit trees, manure and harvest. It was a form of divided property in which the owner might take up to the equivalent of a quarter, or even more, of the harvest in rent. The tenant paid a fee to the landlord when he signed the lease. At the end of the lease, the possibility of expelling the tenant gave the owner a means to pressure him for a higher rent, but the high cost of compensating the tenant for buildings, hedges, ditches, trees and so forth was also a constraint on the landlord; indeed, when such operations took place, they were almost invariably financed not by the landlords but by the new tenants.

Desruisseaux also became a shipowner, taking shares in local coasting vessels along with fellow-investors from Vannes but also with clients from Bordeaux and Nantes. Le Goff studied profits and losses on 26 such coasters and larger vessels between 1737 and 1764, and provides a table of the profits, losses and insurance on 35 grain shipments with the Bordeaux firms of Vignes and Treilhes & Le Chevalier; from the same data, Magali Barthélémy has studied 38 coasters and other ships in the same period.

Was this specialisation planned, or simply a clever tactic in the face of economic uncertainty? The cereal economy had long been directed towards the market. The orientation of its trade towards the coast extending from southwest France into the Mediterranean basin grew stronger during the seventeenth century, a period that Alain Croix has called its *Golden Age* (Croix, 1993: 133) period. And indeed, recurring shortages in the southern market made grain a key product for export from southern Brittany. Grain was the major outbound cargo traded for Mediterranean produce like wine and salt. Grain gained importance as a crop because it allowed the region to fit into the Atlantic economy; it was part and parcel of Breton maritime development. In the long run, from the fifteenth century to the beginning of the nineteenth, grain growing remained essential to agricultural trade. In addition to the opportunities offered by the inter-regional or international market, demand also came from big clients like the East India Company or the Navy, especially in the eighteenth century. During the second half of the nineteenth century, however, a new specialisation developed along the foreshore.

IV. Oyster-farming, a nineteenth-century sea agriculture in the Gulf of the Morbihan

IV.1. Agriculture on the foreshore

This section might seem out of place in a paper on agricultural specialities. Yet oyster-farming can arguably be included in what Jeanine Le Bihan terms sea agriculture (Le Bihan, 2007: 112). Long before the nineteenth century, fishermen-farmers had made oyster-fishing part of managing foreshore resources. This link to both the agricultural and the maritime worlds was permanent. On the one hand the Navy had its reasons to keep up the numbers of registered sailors subject to conscription; on the other, the investors in this new activity – notables, property owners and members of the liberal professions – took over, so to speak, from the landed proprietors who had been involved in the agro-maritime grain system during the previous century. Even though the Navy continued to supervise oyster-fishing, it was the national Agricultural Show of 1875 in Vannes that the pioneers of this new speciality used as the forum to publicise their innovations, drawing up a series of monographs of which 16 examples have been preserved[3]. The vocabulary they used to describe the zone under cultivation and the way the oyster-farms were organized gives ample evidence of the agricultural origins of this branch of activity.

It is no exaggeration to speak of the mid-nineteenth century innovations as a 'turning point'. Jeanine Le Bihan shows that the birth of oyster-farming was closely linked to the scientific progress of the day, but that it was also the fruit of practical innovation and experience. It is remarkable how quickly progress in scientific work on artificial fertilization led to experimentation in the field. But although the results were at first satisfactory, by the 1860s they had failed to live up to expectations. Scientists still did not properly understand the oyster's anatomy and its reproductive process. But on the ground, pioneers built up a body of knowledge by trial and error that gradually allowed them to master the techniques of collecting, breeding, refining and greening oysters. Their work was innovative. Leroux, in Crach, created the definitive method of clustering on tiles, Doctor Gressy of Carnac developed a technique for macadamizing the park beds, Liazard and Martin worked out an original process for liming the tiles. Leroux's observation of the oyster life-cycle made easier to detect the laying season. De Wolboch worked out the best moment to run barges aground to get the collecting boxes to the right place on the beds Detaching the spat required much care. To improve breeding on soils that were not always of good quality, Pozzy developed oyster-friendly cases; and a foreman of hydraulic works at the military dockyard in Lorient invented collector-tubes and 'hives'. When scientists were unable to provide sure methods for refining and greening, the most suitable sites were found by trial and error. Experiments in refining were carried out at Ludré, on the Traict du Croisic, at Pénestin and in the Belon River; the greening process was worked out at Mesquer and Le Croisic. Thus from its outset, oyster-farming involved cooperation between scientific researchers and oyster-farmers with their ongoing practical improvements.

The spread of oyster-farming led to the privatization of the foreshore under State control, in the form of concessions of the maritime public domain, which then made it possible to shape the landscape. The Ordinance of 1681, which defined the maritime domain, was fundamental to this process. The method of concession established in the decree of 1853 gave the oyster-farmers only precarious and revocable tenure, awarded on a non-transmissible, individual basis. The first change to this status came in 1915 with the creation of a 25-year renewable lease with the possibility of an exemption allowing transfer to a third party. But it was the law of 1919, which, by allowing companies to be set up, made long-term concessions possible. In this same law the term ostreiculteur was used for the first time, to designate an 'oyster-farmer',

[3] Archives départementales du Morbihan, 7 M 325.

a term previously used only to designate someone who was not registered for naval conscription.

Existence over the long term is what made it possible to build up this very particular landscape of oyster-beds and worksites, which usually included the platforms, the tender's cabin, the storehouse and the workshop close to the holding basin. Even so, the holder of the concession was required to file an application known as an AOT, that is, a request for a Temporary Occupation Authorisation (*autorisation d'occupation temporaire*).

IV.2. The oyster-farming company: an innovation

Continuous activity along the foreshore of the coastal public domain became a special way in which farming and even more, property transfer, could take place. It still bore the mark of its twin maritime and agricultural origins, brought out clearly by the employment of the expression 'sea agriculture' or 'coastline agriculture'. Oyster-raising was not a single profession but rather a group of professions. It could, to be sure, designate workers at one or several levels of the production cycle such as spat-gatherers, half-breeders, breeders or breeder-refiner-shippers. But certain oyster growers specialized in the various stages of marketing. They did not make up a homogeneous milieu. Put simply, on the one hand, there were the registered sailors for whom the Navy always tried to provide special treatment, even conceding oyster-beds to them within the framework of Sailors' Associations; this was the case in Pénestin and in La Trinité. where they became operators of small oyster-beds occupying less than a single hectare. On the other hand were investors from the local power elites: landowners, doctors and pharmacists, notaries, wholesale merchants and retired customs officers, naval officers or officials of the naval recruitment system.

Contrary to what the legislation said, the oyster-farmer was not always the person who ran the operation: in some cases the holding was farmed out to a tenant. Such people were often termed fictitious oyster-farmers. Because of the seasonal nature of his activity, the oyster-farmer, even when he ran things himself, frequently had a second occupation, often out of necessity. But in reality, the core of the profession consisted of the operating oyster – farmers, who managed to survive over the long term by creating real family strategies, as was the case of the Percevault, Cadoret, and Solhminiac clans. These families knew how to use different legal frameworks available to businessmen so as to make sure their concessions would last. The Martin family of Crach for example, used partnerships and limited liability partnerships. Some large concerns even qualified as public limited – liability companies. But clearly most preferred a simple association of father and son in order to pass the business on in the family. If there were only daughters, it was a son-in-law who had to became the oyster-farmer and acquire the knowledge indispensable to the trade. Marriages between dynasties allowed these people to build up the oyster-farming community. But family labour was not enough. When a tenant did the farming, stewards and managers had to be found. Above all, it was necessary, to recruit a labour force that at certain times of the year could become quite large. In 1876, 2,000 to 4,000 people, depending on the district, were recruited for periods running from six months to a single day. As in the canning industry, a high proportion were women and children. The Navy would have preferred to see this labour force drawn from the body of its registered sailors, combining navigation with work on the oyster-beds; but they were in fact recruited more and more from among the general population of the seacoast. That society was constantly faced with the hardships of a structurally fragile economy. Bad seasons, scarce resources and poor harvests had always characterised the worlds of fishing and agriculture from which oyster-farmers came. Variable weather could generate losses, and so could predators. But it was epizootic disease that left the heaviest mark on the annals of oyster-farming.

Thus, the growing and export of cereal grains drove those involved in the micro-capitalism of Vannes, the merchants and landowners who constituted the local elite (Le Goff, 1972: 26-27), as well as tenant farmers and seafarers. Grain production is central to understanding how southern Brittany fitted into the Atlantic economy from the fifteenth to the eighteenth centuries. But people active in mixed farming also appropriated the resources of the foreshore and shallows, abandoning fishing to go off on their small craft dredging for oysters, along with their women and children. When the oyster-beds ran out in the middle of the nineteenth century, people had to innovate and shift from fishing in order to cultivate them. This earliest form of aquaculture developed under the supervision of the Naval authorities, but the local elite, including landowners and members of the liberal professions, were drawn to support it – the same social groups as those involved in the "cereal system" during the previous century. The new specialisation required new know-how, and new ways of regulating amphibious property concessions. Indeed, the term "concession" itself underlines the problem posed by the legal status of farms located on the inalienable maritime domain. The organisation of oyster-farming and oyster-workers was similar to that of agriculture although certain aspects refer back to its origins in maritime trade or fishing. About the same time, the gathering of sea-fertilizer became part of coastal tradition. In northern and western Brittany, particularly from the mid-nineteenth century onwards, farmers took up a new specialisation, the cultivation of early fruits and vegetables, stimulated by growing demand from French towns, as well as by foreign demand. Nor should we underestimate the contribution made to the local economy by the increasingly popular interest in sea-bathing. English demand for this and other specialized products and services revived cross-Channel relations;

such direct targeting of production for the British market, may without exaggeration be considered a feature of a cross-border economy.

V. Seaside agriculture on the north coasts of Brittany: The enrichment of primary produce with coastal fertiliser

V.1. Sea manure

In 1733, Intendant Les Gallois de La Tour explained that the local economy in the diocese of Tréguier was built around 'good arable land which produces wheat, rye, barley, buckwheat, hay, flax, hemp, and all kinds of vegetables' (Levasseur, 2000: 699). Marlin, who visited Brittany in 1785, likened the Trégor peninsula to a Promised Land (Levasseur, 2000: 705). Le Masson du Parc, drawing the conclusions from his 1726 survey of the littoral for the French Royal Navy, explained why the sea no longer interested the people of the Trégor region and why they preferred agriculture:

> [...] These peasants, who are prosperous and rich, will never leave off cultivating their land to go to the sea; the value of their food products, which is very great in this part of the Lower Breton coast, furnishes them abundantly with all things [...] (Levasseur, 2000: 547).

On Île Grande, Le Masson du Parc noted that small boats were now used to extend their agricultural activity, describing how 'the fishermen and "the inhabitants of this place put to sea' in small craft of a couple of tonneaux' burthen 'to gather wrack and sea sand, which is of greater account than fishing. Peasants come from inland to load both these products of the sea which they use to fertilise their land and particularly to prepare the ground where they are going to sow hemp and flax; this makes the peasants of these districts wealthy and this gives them all the more reason to neglect fishing for ploughing, which they find more profitable' (Levasseur, 2000: 449). Marine fertilisers made possible a trade in agricultural products that earned more than fishing. Le Masson du Parc said as much in speaking of Lannion:

> These products of the sea sell extremely well in Lannion, Fishermen make piles of them during the good weather; they accumulate heaps of sand and sell them to ploughmen who come more or less uninterruptedly, to load these substances, which produce an extremely abundant harvest of hemp, flax, and spring grains. These materials enrich their soil as much as the best manures (Levasseur, 2000: 449).

Shore manure was the great treasure of littoral agriculture. Some time should be taken at this point to clarify the terminology of marine enriching agents. Shoreline wrack or bladderwrack of the Fucus species, which could be got during low tide, and submerged wrack or kelp of the Laminaria species, harvested by boat, are both known as vraic (wrack) in Normandy and sart in Aunis and Poitou. They are what is commonly called 'sea manure'. The only contribution of sea sand to making soil more fertile is breaking down its own molecules. It is the sea salt and bituminous particles it contains that turn it into an excellent fertilizer. Maerls are high quality sand containing fragments of calcified seaweed and shells. In some coastal areas, farmers employ *menusses* (bits of fish and fry) and *guildre* (made up of small shrimp, shells and bits of fish) as fertilizers despite prohibition of the practice.

It is thus abundantly clear that in a coastal region where agriculture dominated, the sea played only a secondary, though essential, role in agricultural production, by providing fertilizer from the shore. On the northern Breton coast, the process continued into the nineteenth century. Aymard de Blois lauded the virtues of maerl to the Agricultural Society in Morlaix and the Conseil Général of the local Departement of Finistère took care to spread the word. In 1825, Jacques Houze extracted maerl at the end of La Frenaye bay and discovered a bar of maerl at Erquy. Soon the whole bay of Saint-Brieuc became a mine for extracting maerl (Denis, Geslin, 2003: 110-111). Many small ports, bays and rias of northern Brittany, such as Lannion, Lézardrieux, Pontrieux, Portrieux, Plancoët and Poulafret, were given a new lease on life by the collection of wrack and maerl. In Paimpol in the year 1862, the trade in sea-fertilizer mobilized a hundred or so small boats usually employed in onshore fishing. Barges drawing fifteen tonneaux were fitted up with a winch to dredge for maerl at depths of 8 to 10 metres. Each dredging brought up one-eighth of a cubic metre. These maritime fertilizers were laid up on the sea-banks where brokers sold them to farmers. Around 1880, a cubic metre of quicksand (tangue) sold for 1.50 francs, and maerl for 2 francs. This was sometimes the only business carried on in small ports and it justified improvint their quays and docks. This was particularly the case in the port of Plancoët, seated on the Arguenon at a distance of nine kilometres from its estuary (Sioc'Han Monnier, 1998: 448). The trade was limited to maerl and quicksand taken from around Le Guildo. There, in 1875, 25 barges discharged more than 50,000 tonnes of these sea fertilizers; farmers from villages up to 40 kilometres away came to the coast to collect them with their carts.

V.2. Early vegetables

Of course, other fertilizers such as bone-black, guano and later, chemical fertilizers were also being used by the end of the nineteenth century. Never the less specialization in vegetable production during this time of revived trade, and particularly cross-Channel trade, across the region known as the 'Golden Belt' depended on this enriched soil. 'The Golden Belt', as Maurice Le Lannou explained, is the whole of the Breton market-gardening region. For no longer than the last hundred years, and often less, across a space stretching back more than five to six kilometres from the shore, the early potatoes from Saint-Malo and the Trégor region, the onions and cabbages of Saint-Brieuc, the cauliflowers and artichokes of Saint-Pol-de-Léon, the strawberries of Plougastel and the peas of Pont-l'Abbé have enriched the hard-working owners of these tiny holdings.

> The sight afforded by these tiny, carefully cherished fields, hemmed in and girded by dry-stone walls running down to the shore along the slenderest of peninsulas, all the way to the little coves where the fishing barks are piled up alongside the fishermen's low houses, forcefully conveys the powerful singularity of Brittany (Le Lannou, 1939: 12).

This new landscape of shoreline market-gardening dates from the 1820s and 1830s, characterized by potatoes around Lézardrieux, Pont l'Abbé and Loctudy, onions and green vegetables at Saint Pol de Léon and Roscoff, strawberries and potatoes at Plougastel-Daoulas. When Marcel Herubel wrote, in 1924 that 'Roscoff is a seaport only because it does market-gardening' (Herubel, 1924: 40), he forgot to mention that at an earlier date, Roscoff had been the port where ships from Liebau in Courland brought the linseed needed to grow the flax of Léon and Trégor. But his statement does point up how the export of vegetables and early fruits revived local port activity, which in turn justified the renewal of port structures. In 1828, a barge left Roscoff fitted out by a local farmer, Henri Olliver, and three associates with a cargo of red onions, bound for the English port of Plymouth. This was the origin of the 'Onion Johnnies' (Menez, 1986: 1-30), a seasonal migration of onion-sellers – some 200 of them in 1860 – who thenceforward peddled their produce around England every year, from July to January or February. On a larger scale, regular connections by steam coasters headed across the Channel or for the port of Le Havre, opened up real opportunities for developing vegetable production. New port installations on the northern coast in the nineteenth and early twentieth centuries were essentially due to the dynamics created by the vegetable trade. As early as 1794, Cambry spoke of Plougastel-Daoulas as 'the garden of Brest and the Landerneau area; there are large strawberry and raspberry plantings, roses and daffodils… Cherry trees, plum trees and apple trees go all the way down to the shore.' The Chilean strawberry, first brought to France by the engineer Frézier, in 1714, in particular to the botanical garden of Brest, adapted perfectly well to the Plougastel peninsula. In 1750, the White Chilean strawberry was blossoming in Plougastel. The success of the strawberry resulted from the combination of two varieties, Chilean and Virginia, to produce a hybrid of exceptional size and red flesh. 'Nowhere in Europe,' wrote J.H. Blanchard, chief gardener of the botanical garden in in the Brest Naval base, in 1878, are strawberries eaten more than in Brest. There are so many during the season that markets, street corners and the entrances of barracks and arsenals are blocked by strawberry merchants (Batissou, 2004: 18-23).

At Plougastel-Daoulas, the introduction of Chilean strawberries in the 1820s started a production that was only marketed locally before 1865. Then the new railway connection opened up new markets, especially in Paris. But the fruit tended to incur damage during transport and costs were prohibitive. So producers turned to the English market, making their shipments from the cove of Lauberlac'h, which constantly required improvement. The gradual spread of short rail lines offered some competition to the small ports, eventually leaving only a few in the trade. Never the less, the nineteenth century had seen the development of an original model of agricultural specialisation based on the high fertility of coastal land and shipment by sea to new markets, especially in England.

V.3. The social context of coastal agricultural development

Using or not using the sea did not in itself make a clear difference between maritime and land-bound populations. Parish registers and public records of births, deaths and marriages, and especially baptismal certificates in some parishes, yield information on those with more than one occupation. They show that a maritime occupation was frequently linked to another that was related to the land. The ploughman-sailor and the farmer-fisherman were typical categories on the coast. Another category was made up of people who built fishing boats and coasters in the rural parishes along the seacoast, the ploughman-builder: men like Jean Le Quer in Carnac. The category of farmer-longboat master was less frequent. In the Aven and Belon region of Brittany, Jean Helloury of Kerdruc en Nevez was a tenant farmer, but also owned a two-tonneau shallop that he used to take the apples, cider, and honey that he produced to the port of Quimper (Boulic, 2005: 66). This combination of cart and shallop was met with much more often on Mediterranean shores. Gilbert Buti provides several portraits of captain-wine growers and sailor-foresters on the Provençal littoral (Buti, 2004: 147-161). But the proportion of multiple occupations mentioned in these certificates should not be overestimated: at best it came to a little more than 3 per cent of the professions declared by the fathers of baptized children as in Sauzon, on Belle-Île (more precisely, 3.33 per cent); and was often less: 1.58 per cent in Carnac and 1.18 per cent in Quiberon. True, institutions did not like to use terms of this kind in their official documents, and preferred standardized categories. So we should not rely solely on self-declaration, but also consider information recorded during the central government's inspections of the seacoast, such as those conducted by the Fisheries Inspector Le Masson Du Parc in 1728. On the Breton sardine shores, the agriculture-fishing combination was carried out by a couple or family. In 1789, The cahier de doléances prepared in Quiberon for the Estates General of 1789 claimed that 'The land is worked only by women, the husbands and children being forced to go to sea for eight months of the year so as to earn their daily bread (Danigo, 1977: 149). A century later, in 1866, before the canning industry changed this balance, an overwhelming proportion of 206 out of 236 Quiberon households included a sailor and a farmer-wife (Le Bihan, 1998: 81). This feminisation of agriculture was particularly evident on the islands of Ushant, Groix and Houat.

Thus, on the littoral, the populations of coastal towns and villages fished and served on inshore shipping, but the plot of land remained a constant. Maritime activity was either a principal or part-time activity. The categories of farmer-fisherman and/or sailor and fisherman-coaster seem to have been the most representative. Moreover, it was the farmer-fisherman or the farmer-coaster who would get involved in commercialmarket gardening. The men of Roscoff, with their red onions, or those of Plougastel, with their strawberries, were all farmer-fishermen. Camille Vallaux (Vallaux, 1906: 173-190) pointed out that there was only one village with a sailors population; the rest were either mixed or farming villages. Plougastel itself was a territory of neither fishermen nor the kind of sailors whom the Navy deemed sufficiently skilled to conscript. The man we witnessed making the pioneer onion shipment from Roscoff to Plymouth in 1828 was a twenty-year-old with a farm on the narrow inlet there. The growing extent of land given over to market gardening was a sign of the success of early vegetables. From Roscoff, vegetable cultivation spread outwards to the neighbouring villages of Plougoulm, Plouénan and Plouescat; Saint-Pol-de-Léon, the main town in the area, became a vegetable market town. The vegetable industry was a microcosm of agricultural society. So far, it has been the 'Johnnies' who have received most of researchers' attention. Starting out as apprentices from 8 to 10 years of age, they became bundlers (making braids of onions) or salesmen for six months of the year in the English ports and towns as part of the 'onion-man' companies under the direction of a boss. There were 200 of them in 1860, 1,273 of them in 1903, and 1,500 in 1929. But when these seasonal workers returned home after New Year, they went back to their small farms. Onion cultivation required plenty of gardening and plenty of workers. Indeed, it was 'broad-bed cultivation, with no part of the ground unused' (Vallaux, 1906: 186). Moreover, when they came back, they had to help with raising potatoes and cauliflowers, before moving on in July to onion-picking in the open fields. All this happened in a region of farms smaller than ten hectares in area, where tenant farming could only be found near Roscoff. It was the poorly-paid day labourers whom Henri Baudrillart (Baudrillart, 1987: 40) described as travelling all over the countryside between Roscoff and Tréguier, renting out their service for the main tasks. The strawberry fields of Plougastel were organised in small squares separated by low dry-stone walls, 'the dirt banks on the sides, set back a maximum of 600 metres from the sea front, their area increasing from 200 hectares in 1875 to 300 hectares in 1879, 570 in 1902 and finally 890 in 1928. Strawberries however, were not a single crop, nor a product reserved only for market gardeners. Strawberry growing alternated with peas, beans and broad beans. The strawberry plots were often surrounded by others where lettuce, cabbage, green beans and peas were grown; in 1908, 180 tons of peas and 400 tons of strawberries were exported from the area. Besides the 500 hectares of strawberries cultivated there in in 1908, there were also 775 hectares of barley, 500 of wheat, 212 of potatoes, 20 of beans, and 15 of peas. By then however, the age of the farmer-fisherman or farmer-coaster was long gone, giving way to that of producers who created professional organizations in 1906 to fight competition more efficiently' (Amghar, 2006: 587-589).

VI. Conclusion

It is tempting to think only of the pressure of markets on agriculture and give short-term demand the dominant role in its transformation. But we should try to see things, as much as we can, from the point of view of rural coastal society. These people knew how to manage their land and tap its potential. In places where farmers knew how to profit from foreshore sources of sea fertilizers, people turned to the sea only as a second choice. Systematic dredging for clay and maerl in the nineteenth century was only a change of scale; for coastal and close-hinterland farmers it was a matter of best adapting a long-practiced activity to growing demand. Rather than insisting on the specialisation of coastal agriculture, we should emphasise its ability to grasp at new opportunities or adapt to new limitations. In the end, the extension of the vineyard at the partial expense of cereal crops, the development of market-gardening, the growing of wheat for speculative markets and the move from oyster-dredging to oyster-farming all show the adaptiveness of a population able to grasp at opportunities or at least to work out ways to live with them. True, we should not overlook changes in types of farming activity and the advantages of a productive soil for introducing new crops, but there were no really great changes in productive organisation or land structure. So innovation or specialisation must have proceeded from optimisation of resources and know-how in rural societies that were already well used to managing their land and time, and able to adapt almost continuously to their environment. In the case of Western France, this was true at least until the end of nineteenth century. Innovation must, therefore, be compatible with some continuity in socio-economic models.

Bibliography

AMGHAR, Julien (2006), *Les Petits ports et les usages du littoral, en Bretagne au XIXe siècle,* unpublished doctoral thesis, University of Bretagne Sud, 1414 p.

BAEHREL, René (1961), *Une Croissance: La Basse Provence rurale (fin XVIe siècle-1789),* Paris, S.E.V.P.E.N., 842 p.

BARTHELEMY, Magali (2002), *Ignace Advisse Desruisseaux, itinéraire d'un négociant Vannetais au XVIIIe siècle,* Master's dissertation, University of Bretagne Sud, 241 p.

BATISSOU, Chloé, 2004, 'Aux origines de la Fraise du Chili à Plougastel', *Armen,* 141, 2004, p. 18-23.

BAUDRILLART, Henri (1987), *Il y a un siècle les paysans bretons,* edited by Fanch Elegoët, Plouguerneau, Tud Ha Bro, 296 p.

BLANCHARD, J.H. (1879), 'Le Fraisier de Plougastel, son histoire, son origine, sa culture, son rendement', *Journal de la Société Centrale d'Horticulture de France,* 39 p.

BOULIC, Ronan (2005), *Les Sociétés littorales du pays de l'Aven et du Belon au XVIIIe siècle,* Masters dissertation, Université de Bretagne Sud, 65 p.

BUTI, Gilbert (2000), ''Le Chemin de la mer' ou le petit cabotage en Provence (XVIIe-XVIIIe siècles)', *Provence historique,* 201, p. 297-320.

BUTI, Gilbert (2004), 'Gens de mer et du terroir: Capitaines-vignerons et marins forestiers de la France méditerranéenne', in Gérard LE BOUEDEC *et al* (eds), *Entre Terre et mer, sociétés littorales et pluriactivités (XVe-XXe siècles),* Rennes, Presses Universitaires de Rennes, p. 147-161.

BUTI, Gilbert (2007), 'Vigne et cabotage en Méditerranée nord occidentale au XVIIIe siècle', in A. CASANOVA (ed.), *La Vigne en Méditerranée occidentale,* 128e congrès national des sociétés historiques et scientifiques, Bastia, 2003, Paris, Éditions du CTHS, p. 9-26.

BUTI, Gilbert (2011), 'Fruits de la rivière de Marseille aux XVIIe et XVIIIe siècles siècles', *Cahier des Annales de Normandie,* 36, Mélanges Zysberg, Musée de Normandie-Université de Caen, p. 87-102.

COLLINS, James B. (1984), 'The Role of Atlantic France in the Baltic Trade: Dutch Traders and Polish Grain at Nantes, 1625-1675', *Journal of European Economic History,* 13, 2, p. 239-289.

CROIX, Alain (1993), *L'Âge d'or de la Bretagne 1532-1675,* Rennes, Éditions Ouest-France Université, 596 p.

DANIGO, Jean (1977), 'Les Doléances maritimes des paroisses de la Bretagne méridionale de la Vilaine à l'Elorn en 1789', *Mémoires de la Société Historique et Archéologique de Bretagne,* t. LIV, p. 135-154.

DENIS, Michel, GESLIN, Claude (2003), *La Bretagne des Blancs et des Bleus,* Rennes, Éditions Ouest-France Universités, 719 p.

DION, Roger (1959) *Histoire de la vigne et du vin en France des origines au* XIXe *siècle,* Paris, Clavreuil, 768 p.

FAUCHET, Joseph (1802), *Statistique générale de la France, Département du Var,* Paris, Imprimerie des Sourds muets, an x.

GAUSSENT, Jean-Claude (1987), 'Un Aspect du commerce maritime au XVIIIe siècle : l'Exportation des vins et eaux de vie par les bâtiments du nord', *Bulletin du Centre des espaces atlantiques,* 3, p. 85-126.

GUILLEMET, Dominique (2000), *Les Îles de l'ouest de Bréhat* à *Oléron du Moyen Âge* à *la Révolution*, Poitiers, Geste éditions, 355 p.

HERUBEL, Marcel-Adolphe (1924), *Le Port de Roscoff, étude d'économie maritime,* Paris, Société d'Éditions géographiques, maritimes et coloniales, 54 p.

LACHIVER, Marcel, (1982), *Vins, vignes et vignerons. Histoire du vignoble français,* Paris, Société historique et archéologique de Pontoise, du Val d'Oise et du Vexin, 957 p.

LE BIHAN, Janine (1998), *La Filière de la pêche de la sardine à Quiberon (1865-1914),* Master's dissertation, University of Bretagne Sud, 105 p.

LE BIHAN, Janine (2007), *Ostréiculture et société ostréicole en Bretagne Sud (1850-1986),* Unpublished doctoral thesis, University of Bretagne Sud, 519 p.

LE GOFF, Tim J.A., (1975), 'An Eighteenth-Century Grain Merchant: Ignace Advisse Desruisseaux', in John F. BOSHER (ed.), *French Government and Society 1500-1850: Essays in Memory of Alfred Cobban,* London, Athlone, p. 92-122, briefly summarized in 'Un marchand de grains vannetais au XVIIIe siècle: Ignace Advisse Desruisseaux', *Bulletin de la société polymathique du Morbihan,* 1972.

LE GOFF, Tim J.A. (1981), *Vannes and its Region. A Study of Town and Country in Eighteenth-Century France*, Oxford, Clarendon [French translation: *Vannes et sa région, ville et campagne dans la France du XVIIIe siècle*, Loudéac, Salmon, 1989, 396 p.].

LE LANNOU, Maurice (1939), *Itinéraires de Bretagne : Guide géographique et touristique,* Paris, J.B. Ballière et fils, 296 p.

LEVASSEUR, Olivier (2000), *Les Usages de la mer dans le Trégor du XVIIIe siècle,* unpublished doctoral thesis, Rennes 2, 1646 p.

LUC, Michel Albert (2005), *Les Gens de mer dans l'île de Ré du XVIIIe siècle (1681-vers 1790), terriens de la mer, marins d'une terre,* unpublished doctoral thesis, University of Poitiers, 749 p.

MENEZ, Jean-Pierre (1986), *Johnnies au pays de Roscoff hier et aujourd'hui,* Rennes, Skol Vreizh, 78 p.

Pinczon du Sel des Monts, (1756), *Considérations sur le commerce de Bretagne,* Rennes, Impr. J. Vatar, 137 p.

Pourchasse, Pierrick (2006), *Le Commerce du Nord, les échanges commerciaux entre la France et l'Europe septentrionale au xviiie siècle*, Rennes, Presses Universitaires de Rennes, 389 p.

Sioc'Han-Monnier, Françoise (1998), *La Construction et l'évolution des ports en Bretagne aux xixe et xxe siècles,* unpublished doctoral thesis, University of Rennes 2, 276 p.

Vallaux, Camille (1906), *La Basse Bretagne, étude de géographie humaine,* Paris Publications de la société Nouvelle de Librairie et d'Édition, 320 p.

Voss, Peter (1995), *Bordeaux et les villes hanséatiques, 1672-1715,* unpublished doctoral thesis, University of Bordeaux, 456 p.

5. A matter of geography? Agricultural specialisation in Southern Sweden (Scania), 1732-1864[1]

Magnus BOHMAN

I. Introduction

Modern day agriculture is characterized by specialized production compared to traditional pre-industrial agriculture where the range of output was varied and the peasant family itself was the producer, processor, distributor, retailer and consumer of agricultural products and also provider of most inputs and support services. The original function of farms, whether isolated or grouped together in villages, was to pursue subsistence agriculture (Mills, 1973). Today, agricultural production is no longer a matter of meeting the essential needs of individual households but rather of producing for the market.

Regional specialisation of production and division of labour is considered a main feature of economic development. Its emergence is therefore often discussed in the context of market expansion and the transfer of capital, goods and labour (de Vries, 1994; Schön, 2000). However, compared to the mechanical processes of industrial production, agricultural production is fundamentally dependent upon biological and natural conditions such as soil type, topography and climate (Cauley 1956: Björkman, 1922). Consequently, the geographic context must not be overlooked when investigating agricultural specialisation.

A thorough investigation of specialisation requires deep knowledge of the actual object of study: agricultural production. Alas, providing relevant estimates of production has proven an elusive task for researchers up to now. This study strives to

1 This work has been done within the projects 'Economic Development and Social Dynamics, Swedish Agricultural Transformation in European Perspective' (financed by the Swedish Research Council) and 'Market Growth and the Spreading of Innovations. New light on the Process and Effects of the Agrarian Revolution' (financed by Handelsbanken : the Foundation of Jan Wallander, Tom Hedelius and Tore Browaldh). I am grateful to fellow project members Mats Olsson and Patrick Svensson. I am also grateful to organizers and participants of the COST-workshop in Rennes 2007 for valuable comments and suggestions.

Agricultural specialisation and rural patterns of development, ed. by Annie ANTOINE, Turnhout, 2016 (*Rural History in Europe*, 12), p. 99-117.

BREPOLS PUBLISHERS DOI 10.1484/M.RURHE-EB.5.112263

overcome these obstacles by employing a new database, which estimates production output of farms on the basis of peasants' annual tithe-payments to the parish priest. (Olsson & Svensson, 2007).

This study covers thirty-four parishes and more than two thousand individual farm households. The sample is representative of regional geography and socioeconomic structure, which makes a satisfactory comparative approach possible. The period studied, 1732-1864, comprises the great transformation of the agrarian society by, among other factors, enclosures and increased commercialization.

This study emphasises the geographic context in identifying and explaining specialisation. We will investigate the impact of factors such as enclosures and commercial incentives but also of socio-economic conditions and farm size. All these explanatory factors will be discussed in the following sections, as well as their transformation into variables for econometric analysis.

II. Agricultural specialisation in Scania

Denmark relinquished Scania to Sweden in the peace of Roskilde (1658). Decades of initial turmoil were to follow but Scania soon settled down to be one of the most prominent and rich agricultural regions of Sweden. Because of its communal nature, the open field system inevitably led to the homogeneous production patterns that prevailed during the eighteenth century. Villages were organised on a similar three-field rotation system, particularly in the plains, where the proportion of cultivated land was high, One gain from having the cultivated fields neighbouring of villages adjacent to each other was that there was less need to enclose crops in order to protect them from livestock. This form of village organization added to the regional homogeneity of production.

Geography affected production and resulted in geographic specialisation, which means that the farmer optimised production to suit local geographic conditions. In practice that could involve for example favouring cereal production in regions enjoying comparative advantages because of factor endowments such as favourable land and weather conditions (Mora & San Juan, 2004). In regions not as favoured by these factors, production could instead focus, on among other options, handicrafts, animal breeding or items connected to regional specific natural resources such as wood (Warde, 2006; Campbell, 1928: 9-17).

This background of geographic specialisation and the organization of agriculture in the open field system helps us better understand the ethnographic division of Scania applied in this study. This classification was introduced by geographer Åke Campbell who proposed roughly homogenous regional production patterns and division of labour that suited regional geographic characteristics such as vegetation, topography and the distribution of natural resources. Briefly put, this ethnographic classification divides the countryside into three types of region and economy: the forest, the plain and intermediate zones. Production in the plains favoured cereal production over other activities, resulting in a corn surplus that covered taxes and other expenditures. The proportion of cultivated land was high but by contrast there was a shortage of fodder and wood; and in order to meet the need for pasturage, animals were often sent to the intermediate areas. Woodland production was in many respects the opposite of that of the plains: diversified and including several specific items, mainly connected to the natural resources of the forest, which paid for taxes and essentials. Less corn was produced than local needs required and the share of cultivated land was low, whereas animal breeding and pasturage were of some importance. In intermediate regions, production and other activities were rather diverse but dominated by pasture and animal breeding, as the soil type did not usually favour arable production. Transportation of various kinds played an important role and there was usually a shortage of both corn and woodland items.

The geographic specialisation suggested by Campbell thus involves two aspects: mostly specialized vegetal production in plain regions and more mixed farming in wood and intermediate land regions. One way to illustrate this economic system, which rests on regional dependence and division of labour, is to think of Scania as one single farm or village. In this metaphor the plains would constitute the arable land, the intermediate regions the pasture land, and the woodland regions the outlying land with the important resources of the forest.

A break with the old order in Scania began with early enclosures, storskiften, that were implemented from the mid-eighteenth century onwards. They might be described as a sort of rationalization process with the aim of minimizing the number of land strips per farm within the open fields of the villages. Eventually a far more radical form of enclosure, enskifte, was introduced in 1803. It resulted in a major reorganisation of agrarian society by abolishing the open field system, shattering villages and establishing private management over individual farm units dotted about the countryside. The process was gradual and proceeded at a different rate of speed on the national level all through the nineteenth and even the early twentieth century in more remote areas. In Scania however, it was mainly completed as early as 1830 (Dahl, 1942). On freehold land enskifte was conducted in response to the individual peasant initiatives. For that reason some researchers have considered it an entrepreneurial initiative by individual peasants for the purpose of establishing private management of the holding and its production (Svensson, 2001: 63). Thus radical enclosures altered the individual farmers' potential to specialize production. We also learn from previous research that enclosures usually promoted corn production, which some ascribed to new cultivation methods and cultivation of new land, especially in remote areas (Gadd, 2000). However, the implementation of the enclosures most likely disturbed production at first because of the major redistribution and reclamation of land it entailed.

The strengthening of private management probably increased awareness of commercial incentives as well, and the possibilities for trading agricultural goods changed radically during the period under study. Since production on the national level seldom surpassed subsistence levels, there were several restrictions on trade at the outset, particularly on the corn exports. Conditions changed gradually and Scania, the first province to produce a surplus, was one of the first few provinces where the restriction against trading in the countryside was abolished in 1775, while the rest of the country only followed five years later. In practice this meant that corn and potatoes could be traded independently of time, place and person (Ahlström, 1974). Other deregulations were to follow. Taxes on all goods transported into cities disappeared in 1810 and in 1825 the national prohibition of corn exports was abolished. Finally the British Corn Laws were repealed in 1846, announcing the start of the 'Free Trade Era' (Fridlizius, 1957). Meanwhile the institutions and logistics of the corn trade gradually improved, for instance through the expansion of ports and storage facilities (Åmark, 1925; Fridlizius, 1957).

We have already discussed how geographic factors such as soil type, vegetation and climate together created geographic contexts or regions that differed noticeably in their production and other activities (Björkman, 1922; Campbell, 1928; Overton, 1996). The regional division of Scania suggested by Campbell is, after all, only a generalization and therefore we believe that differences in geographic conditions such as soil fertility within these regions also had an impact on the strategy of individual farmers. It could be argued that soil fertility is acquired through methods of cultivation and that it changes over time. However, it remains fundamentally dependent upon the type of soil: sand, clay and so on. (Germundsson & Schlyter, 1999: 22-25; Wiklander, 1976). The present-day arrangement of soil types in the landscape has proved quite consistent over time; it dates from the last Ice Age and will endure in large part until the next one, if and when that ever occurs.

The peasantry of Scania was a mixed group. For the purposes of our analysis it can be divided into freeholders (including Crown tenants), and tenants on noble land. In the first case, the peasants owned and managed the land themselves by right of inheritance. In the second case, peasants were tenants on manorial land,

and conditions within this group varied considerably. In some cases, peasant tenants managed their land under a lifetime contract with little influence from the demesne. In others, heavy corvée-type labour duties, combined with restricted access to important resources such as wood, left little room or incentive to peasant tenants to improve their holdings. Another important difference between these two groups was the gain from increasing production and productivity. The amount of profit peasants drew from the surplus they produced probably influenced their responsiveness to commercial incentives. Some farms were directly managed by the demesne, sometimes with a steward or servants living on them. These farms were taxed as separate units but the muscle power, whether in the form of corvée labour or animals, was provided by the demesne. In other words they were not producing to meet the subsistence requirements of a resident family.

The sizes of the farms differed as well. Before the reformation of the tax system, farms in Sweden were taxed according to their mantal. The mantal was a tax unit with medieval roots that initially reflected the production capacity of farms: the higher the mantal, the higher the production capacity. However, since it was a rigid unit and production capacity was not, it lost some relevance over time. Still, estimations show that there was a positive correlation between mantal and farm size in the early 1800s (Olai 1987; Svensson, 2001). The mantal was measured and noted as fractions: one-quarter, one-half and so on.

To sum up and return to the core issue, agricultural specialisation in relation to its geographic context can be understood in two ways. On the one hand, mixed farming occurs where the distribution of natural resources allows various activities or is a risk- spreading strategy, as in the case of specialised vegetal production to hedge against unfavourable conditions,. On the other hand, specialized production occurs where natural conditions favour it, as with vegetal production in areas of soil fertility. However, specialized production presupposes surplus production and the ability to bring in essential goods for trade from other areas. Moreover it is expected that mixed farming will decrease as productivity and commercialization increase. The data section below provides further information on the way different factors are expected to affect specialisation.

III. A new approach to estimating agricultural specialisation

Earlier studies have suggested various innovative methods of estimating specialisation. We might for instance count the numbers of items connected to certain forms of production, such as vessels for dairy-products, using farm inventories as source material. Another more long-term method is to investigate how farm buildings evolved over time to suit certain types of production. Both methods were used by de Vries when investigating specialized production in Friesland between 1550 and 1750 (de Vries, 1975).

By contrast, the present study has the advantage of containing production estimates at the micro level for more than two thousand farms. These production estimates constitute the core of the database used for most of the calculations in this study. The source material allows yearly quantification of the two basic elements of farm production: vegetal production and animal breeding. As in our discussion of specialisation above, we consider evident domination of output by either category to be a sign of clear specialisation of production whereas a balance between vegetal production and animal breeding is an indication of mixed farming.

There are two stages in this analysis. First we use descriptive statistics to identify regional specialisation patterns in different geographic settings. Secondly we perform a regression analysis to study the impact of various factors on specialisation. To incorporate the passage of time in our analysis, we split the data into two periods. The earlier period, 1732-1800, corresponds to the traditional open field system with collective management and a lower degree of commercialization. The later period, 1801-1864, includes the radical enclosures, the emergence of private management and increased commercialization, given even greater force by several trade deregulation measures.

IV. Data and variables

Figure 5.1 below shows the ethnographic regions of Scania according to Campbell and also the parishes of the database. Regions one to three are plains regions, four to eleven intermediate regions, and twelve to fifteen are woodland regions. Statistics regarding intraregional distribution of soil fertility are shown in Table 5.1 below[2].

Table 5.1. Regional conditions of soil fertility

Type of region	*Mean*	*Standard deviation*	*Minimum*	*Maximum*
Plain	7.6	0.8	6.1	8.9
Intermediate	6.0	2.3	3.0	9.7
Wood	5.7	0.8	6.1	8.9

Source. Historical Database of Scanian Agriculture (HDSA).

[2] Soil fertility is measured from poorest to best on a scale from 1 to 10 (an explanation of this classification follows below). Note that the statistics represent only the parishes analysed and not the region as a whole.

Figure 5.1. Scania: ethnographic regions and investigated parishes

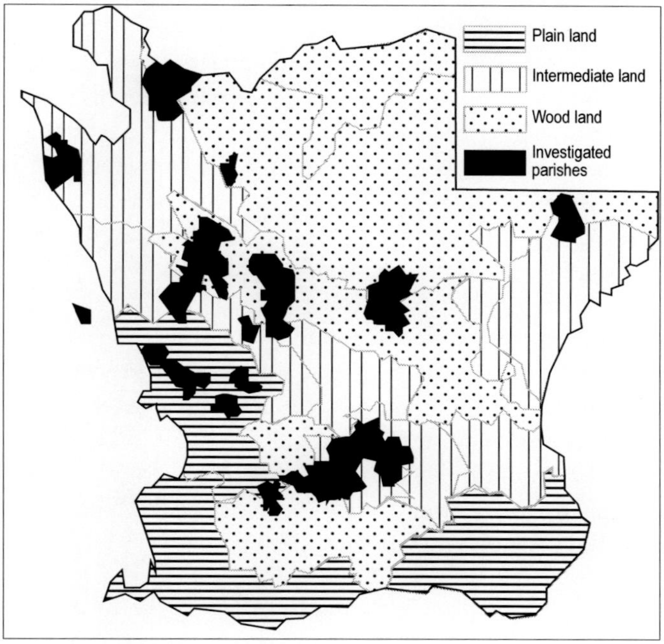

Source. Map by the author. Traditional ethnographic regional classification according to Campbell (1928).

It is possible to estimate production output at farm level from a unique source, based on tithe payments from individual farms to the local parish priest. As this levy remained flexible (*i.e.* dependent upon the annual fluctuations of production) in the parishes studied, we can estimate the total yearly produce of the farms. This particular tithe was claimed as 1/30 of all crops and every tenth animal born on the farms.

Compared to other sources, this tithe has several advantages. In the first place, it was collected at the micro level, and provides extensive information on the production of individual farms. Another, very important, point is its reliability, since the priest collected the levy in person. Moreover, the corn tithes were entered into the accounts as sheaves and not as threshed grain, indicating that it was taken directly from the fields. The period covered the years from 1732 to 1864 but no individual farm is represented consistently throughout the period because there are gaps in the archive material and also because these flexible tithes were converted to fixed amounts at different dates along the way.

The source material had to be adapted somewhat for the purposes of the present investigation. To be begin with, in calculating corn production, a sheaf is neither a weight nor an exact unit of measurement. We have therefore calculated the amount of grain that was threshed, using the personal notes of the priests to estimate the amount of grain obtained from the collected sheaves. This estimate was complemented by material from estate threshing accounts in order to obtain the best possible estimate for different time periods and geographic settings. Previous research suggests that output per sheaf for all crops was almost twice as great in plain regions as in woodland regions (Weibull, 1952), an assumption supported by the threshing accounts. Consequently, the output of sheaves was halved for woodland parishes.

Although potatoes were a new crop of growing importance during the period under study, they are absent from the tithe registers. Since the tithe remained a fixed share of annual production, the parish priests of Scania were to benefit handsomely from increasing agricultural productivity. For that reason a royal ordinance in 1808 stated that they were not entitled to tithes on potatoes. It is however possible to calculate the share of potatoes by using other sources and consequently they are added as a percentage of the total vegetal production for each parish[3].

As we intend to investigate specialisation by studying the balance in production between vegetal products and animal breeding, these two categories have to be made comparable; this is obtained by estimating their respective economic value. The outcome for vegetal production has already been discussed: the economic value of harvested vegetal production is estimated by multiplying the annual yields by the average price for each unit (rye, potatoes and so on) during the investigated period. In practice, this product was consumed, sent to the market or reinvested into production in the form of seeds for sowing, but in the present study we do not consider the relative weight of these outputs.

As to animal production, the source material gives us information on the number of animals (foals, calves, pigs, kids and goslings) born on the farm each year. The economic value of animal production is estimated by multiplying the number of animals born by their respective average price during the investigated period. In doing so, we consider only what the animals' economic value would have been if they had been sold in the market fully grown. In reality, some of these animals were consumed as food, some were used as draught animals (*i.e.* reinvested into production) and yet others sold on the market. Naturally, the livestock also produced output such as eggs and milk and these elements of the economic value of production are not directly

3 For more information about this method and its implementation, see Olsson & Swensson (2007).

estimated. On the other hand, the market value of an animal most likely encapsulates all these items.

Finally, the dependent variable in our analysis is calculated by dividing the economic value of vegetal production by the estimated total economic value of annual farm production (the sum of the economic value of both vegetal production and animal breeding). The produces a quotient, which we will henceforth refer to as the vegetal production quotient (VPQ). Naturally, it varies between zero and one. By multiplying it by 100 we translate it into a percentage. Thus a VPQ equal to 90 indicates that the share of estimated total farm output taken up by vegetal production is 90 per cent and that the share of animal breeding is 10 per cent. Such a case would suggest a farm specialised in vegetal production whereas more equal proportions would suggest mixed farming.

The explanatory variables which are expected to affect specialisation are collected from a wide range of sources, summarized in Table 5.2 below. Each parish in the sample is classified as either plain, intermediate or woodland. Information on soil fertility is summarised at the village level, meaning that all farms within a village are assigned the same value: the average soil fertility of the entire village. Since villages organised in the open field system in practice had their land scattered as strips within the village borders, such a generalization of soil fertility conditions is both necessary and pertinent. Even though it would be possible to estimate the unique conditions of individual enclosed farms after the implementation of the enskifte, such information would require extensive archive work not possible within the frame of this study.

The information on soil fertility is obtained from a national survey of soil production capacity for vegetal production. The classification is the national Swedish standard in modern agriculture and uses a scale from 1 to 10 where 10 is the highest. Based on harvest statistics and soil characteristics, it reflects the production capacity of arable land under normal circumstances (Göransson, 1972). A summary of soil fertility conditions reflects the importance of Scania as an agricultural region since eighty percent of the arable land in Sweden is of quality 1 to 5 whereas the highest grades, 8 to 10, are found only in Scania.

Poll tax registers have been used to provide information on ownership and farms are classified as either crown tenants and freeholders or tenants on manorial land.

Another variable contains information on farm size measured in mantal. Farm management is represented by a dummy variable indicating whether a farm was managed by a peasant or by a demesne. We assume that farm size may affect specialisation so that smaller farms closer to the margins have a higher tendency to be used for mixed farming compared to larger farms producing a consistent surplus for the market. It may also be argued that smaller farms had to maintain a certain minimum degree of animal breeding to satisfy various farm needs, *e.g.* for draught animals. The timing of early enclosures (storskiften), subsequent early enclosures, and radical enclosures (enskiften) is indicated by dummy variables. In addition, yet another dummy variable contains information on solitary farms, that is, farms that were managed outside the open field system and thus already had their land assembled in one unit.

As we have seen, several trade deregulations were carried out during the period under study. They are represented by dummy variables. Nominal prices are available from 1732 and onwards (Jörberg, 1972); relative prices between vegetal products and livestock are calculated as rye prices divided by cow prices. Naturally we expect a positive relation of these variables with the VPQ. Since it is unlikely that relative prices would affect production in the same year, a seven year average was calculated. In other words the variable (see Figure 5.2 below) is expressed as the observation of any given year represented as the average of the seven preceding years.

Figure 5.2. Relative prices of rye and cow, 1738-1860

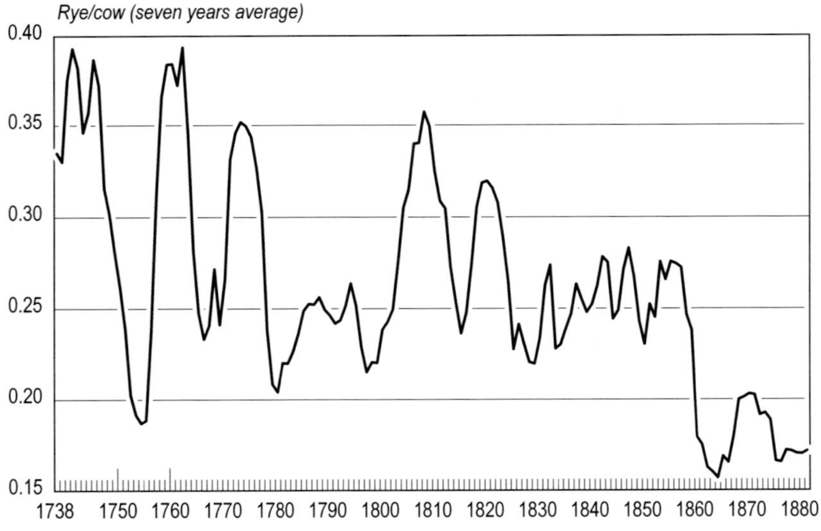

Source. JÖRBERG (1972).

Finally, there were certain circumstances that might be misinterpreted as a sign that peasants were specializing in corn production: the cattle epidemics that ravaged Scania in 1721, 1745-1746, 1749-1750, 1764 and 1768-1769. Cattle mortality was astonishing and in the worst episodes, four out of every five animals died (Weibull, 1923). The timing of the outbreaks is indicated by a dummy variable.

Table 5.2. Descriptive statistics for the entire sample

Variable	*Covariate*	*Per cent of observations*
Type of region	Woodland	43.0
	Intermediate land	32.0
	Plain land	25.0
Type of land	Manorial land	40.0
	Freehold land/Crown land	60.0
Farm management	Peasant1	99.4
	Demesne	0.6
Early enclosures	Not enclosed	66.0
	Enclosed	29.0
	Re-enclosed	5.0
Radical enclosures	Not enclosed	14.0
	Enclosed	77.0
	Solitary unit	9.0
Mean values:		
Farm size	Mantal	mean: 0.29
Soil fertility	Soil fertility (village level)	mean: 6.27

Source. HDSA.

V. Results

First, the general production patterns. In Table 5.3, the mean VPQ for each type of region is summarized as is information for selected representative parishes.

Starting with the mean values for type of region at the bottom of table 3, the assumption of geographic specialisation is borne out, since VPQ is highest in plain land, followed by intermediate and woodland, where production appears more diversified, that is, in areas with the appearances of mixed farming. The pattern is the same if we consider the representative parishes.

There is a positive correlation between vegetal production and soil fertility. However, St. Ibb parish stands out as an exception: its average soil fertility is exceptionally high (9.7) even though its VPQ is low. St. Ibb is however unique in this sample since it is an island (it appears on Figure 5.1, just off the western coast); this may be the reason for its diversified production.

Table 5.3. Mean share of corn, 1711-1860

Parish (type and number of region)	*Mean corn share*	*Standard deviation (mean corn share)*	*Observations*	*Average soil ferertility (10 is best)*
Hög/Kävlinge (2, plain land)	0.79	0.10	1,444	8.9
Brandstad (9, intermediate)	0.66	0.12	2,007	6.9
Tostarp (6, intermediate)	0.64	0.10	498	6.2
Veberöd (15, wood land)	0.58	0.07	2,383	4.8
Hjärnarp (13, wood land)	0.54	0.09	657	5.0
S:t Ibb (7, intermediate)	0.47	0.08	1,667	9.7
Vomb (8 intermediate)	0.41	0.09	1,556	3.4
Billinge/Röstånga (14, wood land	0.38	0.09	6,331	5.8
Mean plain	0.79			
Mean intermediate	0.55			
Mean wood	0.50			

Source. HDSA.

The results presented so far leave out important information such as longitudinal variations and trends. Table 5.4 below presents the summarised results split into two periods, giving vegetal production shares for the earlier (1732-1800) and later (1801-1864) periods. Figures 5.3 and 5.4 show the mean corn share for the representative parishes as graphs. The early period comprises the traditional open field agriculture and the late period includes the radical enclosures and the emergence of private management and also increased commercialization accentuated by trade deregulation.

Table 5.4. Mean corn share, early and late period

Parish (number and type of region)	*Early period: 1711-1800*	*Late period: 1801-1860*
	Mean corn share	*Mean corn share*
Tostarp (6, int.)	no data	0.64
Hjärnarp (13, wood)	no data	0.54
Brandstad (9, int.)	0.61	0.77
Hög/Kävlinge (2, plain)	0.79	0.83
Veberöd (15, wood)	0.56	0.62
S:t Ibb (7, int.)	0.41	0.52
Vomb (8, int.)	0.42	0.41
Billinge/Röstånga (14, wood)	0.35	0.43

Source. HDSA.

Figure 5.3. Mean corn share.
Hög (plain land) and Billinge (woodland) parishes

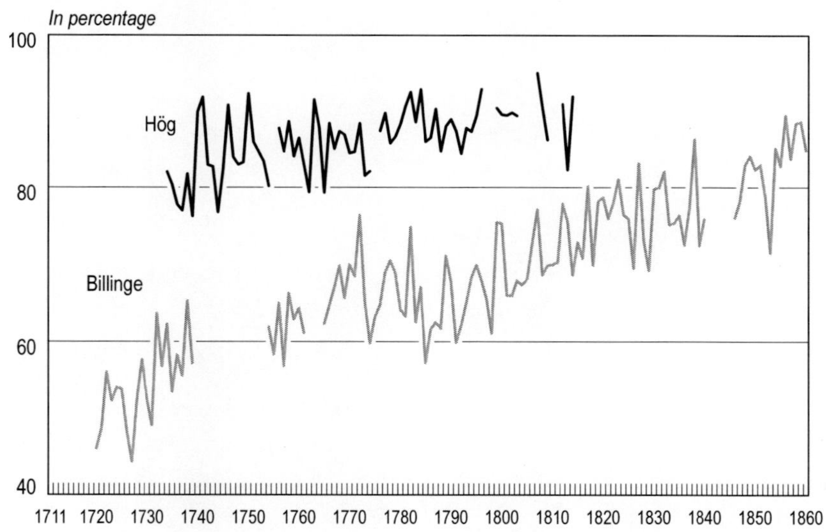

Source. HDSA.

Figure 5.4. Mean corn share.
Veberöd and Brandstad (intermediate land) parishes

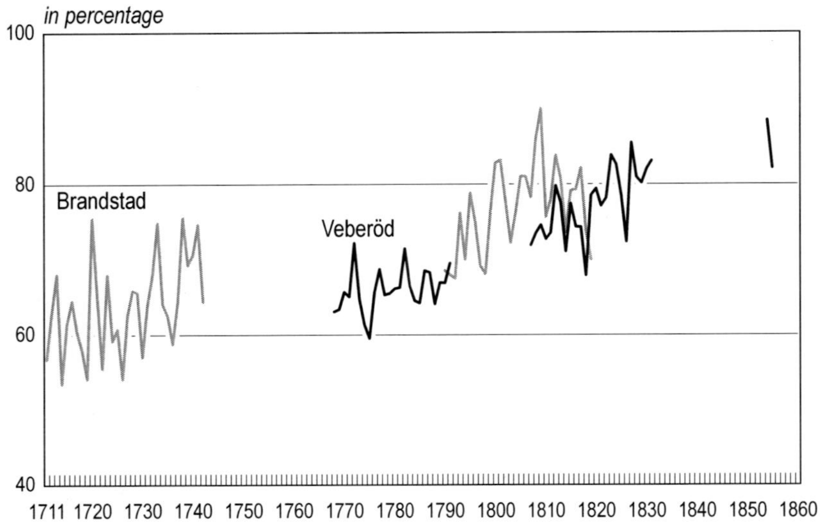

Source. HDSA.

Table 5.4 and Figures 5.3-5.4 show that VPQ increased in all the representative parishes over time. Most notable in Table 2.4 is the increase in Brandstad parish, greater than 12 per cent. The smallest increase is found in Hög, a parish in the plains, which is not surprising since the VPQ was initially very high, suggesting an early specialisation in corn production. Still, differences between regions persisted over time and plain land parishes usually had a higher VPQ than intermediate and woodland parishes.

A more thorough investigation of the impact of our explanatory variables on our dependent variable, the vegetal production share, requires a regression analysis and the panel data set is well suited to using a general least squares (GLS) regression. The results appear in Table 5.5.

The value of the overall r-square value shows that the independent variables explain 25 per cent of the variations in the dependent variable. The significant negative coefficients for type of region imply that vegetal production contributes around 8 per cent less to the total output of a woodland farm compared to a plain land farm and 7 per cent less for a farm in the intermediate areas.

As for soil fertility, the coefficient is both positive and significant, supporting our assumption that local variations in soil fertility have a considerable effect on specialisation in vegetal production. In practice one step higher in soil fertility on the scale from 1 to 10 is equal to 1.42 per cent in VPQ. Considering the minimum and maximum values of the sample (3 and 9.7), the village with the best soil fertility consequently has a VPQ that is 9.5 per cent higher than the least fertile village[4].

Farm size has a positive and significant coefficient, meaning that the bigger the farm the higher the VPQ. One explanation would be that a small farm has to maintain a minimal number of draught animals, sufficient to manage the agriculture of a significantly larger farm. Moreover, a small farm is probably closer to subsistence levels and therefore inclined towards mixed farming, whereas a large farm is more prone to specialize, *e.g.* by producing a surplus of corn for the market.

No significant difference can be detected between freeholders and crown tenants on the one hand and tenants of manorial land, on the other. Even if the coefficient were significant, it would still be much too small to suggest that freeholders were more likely to devote themselves to specialized vegetal production than tenants on manorial land.

The coefficient of demesne management is both evident and significant, meaning that farms managed by an estate have a VPQ 6.53 per cent higher than farms managed by peasants. The probable explanation according to our expectations is that farms under demesne management did little or no animal breeding, since muscle power was provided by the demesne.

As for relative prices, the coefficient is significant and positive. It implies that peasants responded to price incentives, in our case by expanding corn production when prices had been favourable for a long period of time. To interpret the effect we can compare Figure 5.2 with the series of relative prices and examine one of the upward trends, for instance during the period 1754-1762. The difference between these years (0.39 - 0.19 = 0.2) is multiplied by the coefficient (10.94 x 0.2 = 2.2). The resulting value tells us that the increasing relative price during the period caused the VPQ to rise by 2.2 per cent.

The cattle epidemics had, as expected, a significant and positive coefficient and effect. The effect would probably be more dramatic effect if only calves were considered, rather than including all animal categories, as in the present case.

---

[4] This effect is obtained by calculating the difference in soil fertility (9.7 - 3 = 6.7) and multiplying it by the coefficient for soil fertility (6.7 x 1.42 = 9.5).

Naturally, the effect has nothing to do with specialisation but rather results from a highly unwelcome cattle plague.

Table 5.5. Determinants of specialisation, GLS regression

Corn share	*Coefficient*	*Standard error*	*P>[z]*
Plain land Intemedate land	r.c. -7.73	0.47	0.00
Woodland	-7.98	0.56	0.00
Soil fertility	1.44	0.11	0.00
Relative price	12.67	1.12	0.00
Mantal (farm size)	12.29	0.97	0.00
Tenants (manorial land)	r.c.		
Freeholders and Crown tenants	-0.13	0.32	0.68
Peasant management	r.c.		
Managed by demesne	5.56	0.99	0.00
Not enclosed (open field)	r.c.		
Early enclosure	-0.45	0.22	0.04
Repeated early enclosure	0.74	0.45	0.10
Not enclosed (open field)	r.c.		
Enclosed (radical enclosure)	1.54	0.22	0.00
Solitary	-2.76	0.71	0.00
Trade dereg. 1775	3.67	0.18	0.00
Trade dereg. 1810	4.68	0.14	0.00
Trade dereg. 1825	3.14	0.17	0.00
Trade dereg. 1846	5.40	0.27	0.00
Cattle epidemics	3.28	0.28	0.00
Constant	61.96	1.06	0.00
Number of observations	45,443		
Number of groups	1,693		
R-sq: within	0.0979		
between	0.4610		
overall	0.2757		
Wald chi2 (16)	6,431.17		
Prob > chi2	0.0000		

Note. r.c. denotes reference category.

VI. Conclusions

Scania, known as 'the granary of Sweden', became more commercialized during the period studied here. The results show there was a great increase in vegetal production relative to animal breeding in all the parishes studied, that is, a transition from mixed farming to specialised vegetal production. However, regional production patterns persisted, suggesting that specialisation remained dependent on geographic conditions.

During the earlier period (1732-1800), farms appear to have been more devoted to mixed farming than in the later period (1801-1864) but parishes in the plains were already in an early stage of specialisation in vegetal production. The persisting differences between regions speak in favour of a clear connection between production and geography, supporting the ethnographic division into wood, intermediate and plains regions and implying geographic specialisation of production. Over time, vegetal production increased in relation to animal production in all the parishes under study. Trade deregulation measures and price incentives stand out as main explanations for this transition from mixed farming to more specialized vegetal production, Regional type and local variations in soil fertility were also important explanations for this process.

To summarise, the increased importance of vegetal production during the period under study most likely reflects a path dependency in Scania as well as efficiency in a context of expanding national and international markets for corn when institutions and logistics were optimised for this type of production. After all, Scania, at least on its plains, possessed geographic conditions favourable to vegetal production that made it the first province in Sweden to produce surpluses. This initial comparative advantage probably played an important role when the Swedish economy as a whole grew more commercialized and regional specialisation took shape on the national level.

Sources

Historical Database of Scanian Agriculture (HDSA), *Department of Economic History,* Lund University:

Enclosure Acts, *Lantmäteriet*

Land Surveyors' accounts, *Lantmäteriet.*

Poll tax registers, *Regional Archives of Lund.*

Tithe rolls, *Church Archives,* Regional Archives of Lund

Bibliography

AHLSTRÖM, Göran (1974), *Studier i Svensk ekonomisk politik och prisutveckling 1776-1802,* Lund Studentlitteratur.

ÅMARK, Karl (1915), *Spannmålshandel och spannmålspolitik i Sverige 1719-1830,* Stockholm, Nord, bokh.

BJÖRKMAN, Thure (1922), *Sveriges naturliga jordbruksområden. Meddelanden från Lunds universitets geografiska institution,* Ser. B.

CAULEY, Troy J. (1956), *Agriculture in an Industrial Economy,* New York, Bookman Associates.

CAMPBELL, Åke (1928), *Skånska bygder under förra hälften av 1700-talet,* Uppsala, A.-B. Lundequistiska Bokhandeln, 279 p.

DAHL, Sven (1942), *Storskiftets och enskiftets genomförande i Skåne,* Scandia –tidskrift för historisk forskning, Band XIV.

DE Vries, Jan (1975), 'Peasant Demand Patterns and Economic Development: Friesland, 1550-1750', in W.N. PARKER, E.L. JONES (eds), *European Peasants and Their Markets. Essays in Agrarian Economic History,* Princeton, Princeton University Press, p. 205-266.

DE VRIES, Jan (1994), 'The Industrial Revolution and the Industrious Revolution', *The journal of Economic History,* 54, 2, p. 249-270.

FRIDLIZIUS Gunnar (1957), *Swedish Corn Export in the Free Trade Era,* Lund, Gleerups förlag.

GADD, Carl-Johan (2000), *Den agrara revolutionen, 1700-1850. Det svenska jordbrukets historia, del 3: Natur och Kultur,* Stockholm, LT:s förlag.

GERMUNDSSON, Thomas, SCHLYTER, Peter (1999), *Sveriges Nationalatlas: Atlas över Skåne,* Uppsala, Almqvist & Wiksell.

GÖRANSSON, Gert (1972), *Gradering av åkermark kring tätorterna i sydvästra Skåne,* Meddelanden om samhällsplanering i Malmöhus län, 3, Länsstyrelsen planeringskansliet.

JÖRBERG, Lennart (1972), *A History of Prices in Sweden 1732-1914,* Lund, Berlingska Boktryckeriet.

MILLS, Dennis R. (1973), *English Rural Gommunities: The Impact of a Specialised Economy,* London, The MacMillan Press.

MORA, Ricardo, SAN JUAN, Carlos (2004), *Geographic Specialisation in Spanish Agriculture before and after Integration in the European Union,* Regional Science and Urban Economics, 34.

OLAI, Birgitta (1987), *'...till vinnande av ett redigt storskifte', En komparativ studie av storskiftet i fem härader,* Uppsala, Almqvist & Wiksell International, 130 p..

OVERTON, Mark (1996), *Agricultural Revolution in England. The Transformation of the Agrarian Economy 1500-1850,* Cambridge, Cambridge University Press.

OLSSON, Mats, SVENSSON, Patrick (2007), 'Agricultural Production in Southern Sweden 1700-1860 – Estimates and Explanations', Presented at the Seventh Conference of the European Historical Economics Society, Lund, Sweden, 29 June-1 July, 2007.

SCHÖN, Lennart (2000), *En modern svensk historia. Tillväxt och omvandling under två sekel,* Stockholm, SNS Förlag, 45 p.

SVENSSON, Patrick (2001), *Agrara entreprenörer. Böndernas roll i omvandlingen av jordbruket i Skåne 1800-1870,* Lund, Almqvist & Wiksell International, 2001.

WARDE, Paul (2006), *Ecology, Economy and State Formation in Early Modern Germany,* Cambridge, Cambridge University Press, 320 p.

WEIBULL, Jörgen (1952), *Tionden i Skåne under senare delen av 1600-talet,* Lund, Gleerups förlag.

WEIBULL, Carl Gustaf (1923), *Skånska jordbrukets historia,* Lund, Gleerup.

WIKLANDER, Lambert (1976), *Marklära,* Uppsala, SLU Service/Repro.

II
Disruptions and their impact on specialisation

6. In transition from a 'traditional' to a 'specialised' agriculture: farming in the Duchy of Brunswick between the eighteenth and the twentieth century[1]

Andreas KULHAWY

I. Introduction. Agricultural conditions and land tenure

The duchy of Brunswick was situated in northern Germany not very far from Hanover. During the nineteenth century, the whole country covered an area of 3,690 square kilometres (Meibeyer, 2000: 23; Buerstenbinder 1883: 80). In 1793 there were some 180,000 inhabitants (Albrecht, 2000: 584). By 1900, as a result of industrialisation and urbanisation, this figure rose to about 460,000 (Schildt, 2000: 804).

The country contained three main regions. The first area surrounded the capital, Brunswick, and included the smaller towns of Helmstedt and Wolfenbüttel. It consisted of quite fertile lowland and rolling countryside. The second area lay by the river Weser and included fertile land on the banks of the river and less fertile areas in the hilly country above the Weser. The third part lay in the Harz Mountains, where soil and weather conditions made the land quite infertile. In the Harz and Weser area, agriculture was mainly carried on as a subsidiary to mining and forestry, with cattle breeding predominating. Farming was mainly directed towards self-sufficiency. This situation did not change very much until the middle of the twentieth century. For this reason we will concentrate on the fertile area around the capital.

In the eighteenth and nineteenth centuries, agriculture in the duchy of Brunswick was dominated by small and medium-sized farms ranging from a few hectares

[1] The following essay is essentially based on the first major part of my PhD-thesis on the Brunswick Leihhaus as an instrument of modernisation ('Das Braunschweigische Leihhaus als Instrument der Modernisierung. 1832-1914') which is submitted to the faculty iv for Human- and Social sciences of the Carl-von-Ossietzky-University Oldenburg. This paper is an amended and enlarged version of another paper published in Prager wirtschafts- und sozialhistorische Mitteilungen, 15, 2012, p. 28-40.

Agricultural specialisation and rural patterns of development, ed. by Annie ANTOINE, Turnhout, 2016 (*Rural History in Europe*, 12), p. 121-134.

BREPOLS PUBLISHERS DOI 10.1484/M.RURHE-EB.5.112264

to a maximum of about 50 hectares in surface. The average size of an estate was 8.6 hectares (Abel, 1990: 135). There were few noble estates. At the end of the eighteenth century between 75 and 85 per cent of the arable land was in the hands of farmers. The nobility cultivated only 7 per cent by itself. The remaining land was state property or owned by ecclesiastical institutions. Farm ownership was split between peasants and lords, as in most parts of Germany west of the river Elbe. Under normal conditions, the peasants could not be evicted from their farms (Schneider and Seedorf, 1989: 26; Wittich, 1896).

This divided ownership of the farms as an aspect of feudal society, had developed differently in most of central Europe since the middle ages. It originated in Roman and Teutonic ideas of law; Italian jurisprudence coined the expressions dominium utile and dominium directum for it. In the duchy of Brunswick the divided ownership of the farms took its final form quite early, in 1597, when farms became practically hereditary. The opportunities for landlords to evict peasants from their farms were strictly limited. The object of these measures was to prevent the landlords from driving the peasantry off their land, because the peasants were the mainstay of the duchy's public finances.

The co-ownership or superior-ownership of the lords required the peasants to pay several dues. The most important of these were the tithe, services to the lord and annual charges on the farms. The dues were usually paid in natural produce. This gave the lords a say in what the peasants were cultivating. As the lords had to sell these goods on the market, they preferred wheat. Thus, just as in most of the other parts of Germany, true market-oriented specialisation of farming was obstructed. However, the traditional system of agriculture caused several more problems. The tithe was taken out of the farmers' gross product. This discouraged them from trying to raise yields. Moreover, the services owed to the lords forced the peasants to keep more equipment in store than they needed for their own estates.

Another problem that held back specialized cultivation was the scattered location of the peasants' land which was often inaccessible except by crossing the property of other farmers. This meant that they were forced to co-operate and to cultivate the same crops. Furthermore, because of the joint ownership of the common land, it was used quite ineffectively. The bonds between peasants and lords and between the peasants themselves, as well as the divided ownership of the land, hampered modernisation, agricultural specialisation and the increase of production.

By the end of the eighteenth century, in the duchy of Brunswick, about 90 per cent of the arable farming consisted of cereals, chiefly rye, which the farmers needed for their own subsistence (Abel, 1990: 136). As wheat-growing was geared towards the market, this kind of production may be described as partially and compulsorily specialized.

II. Modernization efforts in Brunswick agriculture during the eighteenth and nineteenth centuries

In the second half of the eighteenth century most German governments, including that of the Duchy of Brunswick, made efforts to increase agricultural production (Abel, 1978; Achilles, 1978; Achilles, 1993). They tried to emulate English agricultural modernisation in order to strengthen the economic power of the country, but without changing the social and political system. They started by sharing out the common land (Brakensiek, 1991; Albrecht, 1980, 209) and then proceeded to convert services owed to the ducal administration into annual payments. In several cases, even payments in kind were changed into annually paid money equivalents. By these measures, the government tried to help the farmers rationalise the running of their estates, expand their acreage and enter the currency-based market economy. The administration also tried to persuade the farmers to use new methods of cultivation. They were advised to keep their cattle in cowsheds throughout the year in order to obtain more natural fertilizer from them, and to use marl as a primary mineral fertilizer (Venturini, 1847: 23, ff.). Furthermore, farmers were taught to cultivate peas, beans, vetches, flax, cabbage, turnips or rape-seed in the fallow land (Brakensiek, 1991: 218).

The success of these attempts at modernizing cultivation was limited. Food production rose only by expanding the acreage into the common land. The average yield per hectare did not really increase until the second half of the nineteenth century. The cultivation of cereals and production for self-sufficiency continued to dominate. Any specialisation could only be an incidental result. Three explanations can be offered for this lack of improvement, Firstly, the ducal government did not remove the bonds between the peasants and the lords and between the peasants themselves. Secondly, theoretical and practical knowledge about agriculture and how to organise agricultural reforms was insufficient. Thirdly, the ducal administration was not properly trained in implementing the reforms that could make the peasants more willing to carry out specialisation.

In the duchy of Brunswick, further attempts at modernisation ended after 1780. Because of the bad economic situation at that time, reforms under the French occupation government between 1807 and 1813, aimed at loosening the bonds between peasants and lords, were not very successful. This situation lasted until 1830, when the duchy government revived efforts to modernize agriculture. The government, reflecting political developments after the French revolution, adopted a more liberal philosophy, seeking to bolster the political system with a free and land-owning farming community. In this way, the developments of British agriculture were implemented. These objectives meant that the bonds between peasants and lords had to be redeemed by paying money equivalents instead of feudal dues. The money the farmers needed was lent by a state-owned bank (Kulhawy, 2007). Many of the duchy farmers paid interest and repayment annually for the first time. Thus, the peasants were really forced into the money-based market-economy and directed towards making profits. These measures were supported by an upturn in the agricultural economy, starting in the middle of the 1830s, which made it less difficult for the peasants to pay the redemptions. The average duration of repayment was 28.4 years. In most cases, the peasants finished the repayment between 1860 and 1890, however in a very few cases it lasted until the 1920s.

In addition to loosening the bonds between peasants and lords, the bonds between the peasants themselves were also removed. The common land was shared out, new roads and tracks were built, the agricultural land was reallocated and, last but not least, basic improvement of the soil was carried out. These measures greatly improved the agrarian structure. Now any farmer could easily get to his land. These structural reforms also provided the opportunity to intensify cultivation and extend acreage. Together with the removal of the bonds between the peasants and the lords, farmers gained the opportunity to make independent decisions about which crops to grow and how to cultivate their land.

Despite the shift of the land into undivided farmer ownership, the peasants did not receive full power to dispose of their estates. In order to preserve the agricultural structure of small- and medium-sized estates, the division of the farms as well as the consolidation of farms by inheritance or purchase was forbidden. The possibility of selling a farm depended on state agreement.

III. Other measures to promote the Brunswick agriculture

At the end of the 1830s, the duchy started to build a state railway system (Siebenbrodt, 1938) which, by the end of the 1840s, was integrated into the first east-west German railway line, connecting the Prussian capital Berlin with the newly industrializing region along the Rhine and Ruhr rivers. Although the state organised the railway system, it transferred the responsibility for building country roads to the villages and local administrations (Müller, 2000: 194, 225). This gave the peasants the chance to connect into the new traffic networks. As a result, the road network expanded enormously and the duchy's farmers were integrated into the national market. This was a most important incentive for specializing agricultural production in the fertile region around the ducal capital.

Additional support for the development of a modern and specialized agriculture in the duchy came with the founding of an agricultural club in 1832. (Buerstenbinder, 1881: 71, ff.; Pommer, 1912: 351, ff.). Initiated by the governing elite and partly integrated into the administration, it started with just 69 members. Very soon it was opened to ordinary farmers and by 1880 the club had 2,462 members. At that time 24,000 estates in the duchy farmed arable land. Half of them were smaller than 5 hectares and many of their owners probably were not members of the association. The main objective of the club was to represent farmers' interests and teach them modern methods of cultivation and bookkeeping.

The club used various devices to spread modern methods of agriculture among farmers. From the outset, the association published a periodical and brochures on farming which were distributed in the villages of the duchy. From 1868 the club employed an agricultural instructor who travelled through the villages teaching the peasants modern farming techniques. In 1869 the club founded an agricultural school in the small town of Helmstedt where farmers' sons were taught modern methods as well as book-keeping. By 1900 the school grew to become the biggest of its kind in Germany, with about 600 pupils in 1912. The club organized agricultural exhibitions in various localities, displaying the most up-to-date agricultural tools, the best seeds, the most productive cattle and fertilizers. On these occasions, prizes were awarded for the best agricultural products. Last but not least, from 1862 the club operated research fields and an agricultural research institution to assess tools, fertilisers and seeds

From the 1850s, the club promoted the growing of sugar beet and intensified year-round stabling. From the mid-1870s it also pushed to set up agricultural dairy co-operatives and co-operatives for buying seeds and fertiliser. In the early twentieth century, it played a role in founding co-operative agricultural banks.

IV. The development of the German sugar industry

As a result of the agricultural reforms, the improved economic climate and the development of the new traffic networks, agriculture in the fertile regions of the duchy improved greatly. The most important change was the cultivation of sugar beet at the end of the 1840s. From the 1860s this development was supported by the customs duty policy of the German Zollverein (tariff union) and the German Empire.

Since the end of the sixteenth century, sugar consumed in Germany had been produced from imported crude sugar made of Caribbean sugar cane. The origin of the modern German sugar industry lay in the second half of the eighteenth century (Ellerbrock, 1993: 113, ff.; Müller, 1979: 9, ff.), when in 1747, Andreas Sigismund Margraf discovered sugar in the mangel-wurzel. From 1786, Franz Carl Achard experimented with sugar made from plants and in 1801-1802 he established the first sugar plant to use sugar beets. From 1806 the development of the sugar industry was encouraged by the Napoleonic Continental System. But when the Continental System collapsed, German sugar plants were shut down until 1822 (Krawinkel, 1994: 13). Sugar from sugar cane underwent a renaissance, as it was much cheaper at that time.

However, only ten years later, in 1831, the merchants Louis and Carl Hanewald and the pharmacist Conradin Friedrich Eduard Zier founded a model sugar plant in the small town of Quedlinburg, not very far from Magdeburg (Müller, 1979: 14; Krawinkel, 1994: 19). This processing plant was influenced by French examples. The production of sugar from sugar beet and the foundation of new sugar processing plants greatly increased in 1835, when customs duties on sugar in the Zollverein rose. By 1840, 145 sugar processing plants had been founded in the Zollverein. But when the revenues of customs duties shrank because of this domestic sugar production, German governments brought in taxes on the quantity of sugar beets used in sugar production (Krawinkel, 1994: 180). These taxes were increased in the ensuing decades. From 1853, the fiscal burden on sugar from sugar beets was approximately the same as that placed on cane sugar (Ellerbrock, 1993: 232).

Yet taxes were not a lasting problem for the development of the German sugar industry. The fiscal constraint encouraged an ongoing process of modernisation and expansion in the industry, as the tax-burden went down if the yield of sugar per unit of weight rose. Thus after a short depression following the introduction of sugar taxes, the number of sugar processing plants rose constantly, peaking in 1890 with 406 establishments.

In the long run, sugar from sugar beet completely replaced sugar from sugar cane. Already by 1850, sugar production from sugar beets exceeded that from sugar cane (Krawinkel, 1994: 45; Ellerbrock, 1993: 220). By 1895-1896 the share of sugar from sugar cane dropped to under 0.1 per cent. This development had not been the avowed goal of the policy of the German states, but it is clear that tax and customs policy had greatly promoted the sugar-beet-industry (Krawinkel, 1994: 182, 243). All in all *per capita* sugar consumption multiplied as sugar prices fell. In 1836-1837 the annual *per capita* sugar consumption was 2.04 kg; this figure rose to 12.34 kg by 1895-1896 and to 20.01 kg by 1913. Supply and demand stimulated one another.

The centre of German sugar production was in the Prussian province of Saxony, especially in the Magdeburg region, directly adjoining the duchy of Brunswick. In 1841-1842 the province produced 52.5 per cent of the German sugar output (Ellerbrock, 1993: 208, 401). This figure rose to 67.3 per cent in 1860- 1861. The share grown by the province subsequently fell to 46.7 per cent in 1875- 1876 and to 26 per cent in 1907-1908, as other parts of Germany that had not produced much sugar before also turned to sugar production: the Prussian territories of West and East Prussia, Pomerania and Posen and the grand duchy of Mecklenburg. The share of traditional sugar-producing German territories like Anhalt and the duchy of Brunswick decreased at that time.

In the early decades of the sugar industry, German sugar processing plants were usually founded by merchants, noble and civil landowners, higher public servants, entrepreneurs and capital owners, just as in other European countries like Spain (Müller, 1979: 32, ff.; Ellerbrock, 1993: 215, ff., 301, ff.; Martin-Rodriguez, 1982: 360, ff.; Pinilla, 1995: 233, ff.). Rural joint-owners of sugar plants were quite rare in the early years. The sugar beets that were needed had been bought to a large extent from farmers and in most German regions the sugar industry stimulated the farmers to cultivate sugar beets (Krawinkel, 1994: 46). However, in later decades it became more and more difficult for German sugar plants to obtain enough sugar beet of the quality required. Because of this, from the 1860s, sugar plant companies admitted more and more farmers as shareholders. In return every shareholder had to guarantee delivery of a certain quantity of sugar beets per share of the required quality. From then on, more and more sugar plants came to be owned by farmers, especially in the Prussian province of Hannover. This did not occur in other European countries.

V. The Brunswick sugar beet economy

The whole of the Brunswick and Magdeburg region had the ideal soil and climate for growing sugar beets (Müller, 1979: 22, ff.). The sugar industry of this region also benefited from the early rise of the railways, the highly developed road network and, in the case of Magdeburg, the river Elbe. These linked the area to its markets and a large reserve army of farmhands. Moreover, fuel needs were satisfied by the brown coal mines of the region. Last but not least, the agriculture of the Magdeburg and Brunswick area had been intensified since the eighteenth century, so that peasants were familiar with the agricultural techniques they needed when the sugar beet began to be grown.

Contrary to what happened in the other German states, and other European regions like those of Zaragoza and Granada in Spain, the main characteristic of the Brunswick sugar industry was that from its very outset, the farmers themselves owned the sugar plants, which were organized as public limited companies. Among the shareholders of these sugar plants were not only the owners of the bigger farms but also those of medium-sized estates. As mentioned above, the farmers usually had to promise to cultivate a certain number of hectares in sugar beet for every share they owned. Thus, the difference between the owners of medium and big farms was the number of shares they could afford to own. In 1882-1883 the duchy of Brunswick had the second-highest average of shareholders per sugar plant (121) after the province of Hannover (154) (Müller, 1979: 37). The first sugar plant of this kind was founded in 1849 at the small village of Üfingen. By 1850, 5 such sugar plants already existed; by 1880 the number had risen to 26 plus 5 sugar refineries (Buerstenbinder, 1881: 131).

Although the duchy was one of the smallest German states it soon became one of the five most important sugar-producing areas (Ellerbrock, 1993: 208, 401). During the sugar-production year 1840-1841 the duchy produced no sugar at all. In the period 1850-1851 production in the duchy amounted to 2.5 per cent of German output, putting it in fifth place among the sugar-producing German states. In 1860-1861 this share rose to 5.4 per cent and fourth place. It peaked in 1875-1876 with a maximum share of 9.2 per cent and third place. In the following years the duchy's share fell back to fifth place again with 5.5 per cent in 1907-1908.

VI. The role of the Brunswick farmers in the development of the domestic sugar industry

Participation in the process of specialisation was obviously no problem to the farmers of the duchy of Brunswick even if they were still repaying their redemption loans. The sugar plants in the two villages of Vechelde (with a sugar plant was founded 1857) and Wierthe (with a plant dating from 1865) – neither of them very far from the duchy's capital Brunswick – provide good evidence of this. The investment of individual farmers ranged from several hundred to around 4,000 thalers. These amounts in most cases clearly exceeded the balances owed on the redemption loans. All in all, until the middle of the 1860s, 101 peasants with medium and large estates became shareholders of these sugar plants. Sixty-two of these farmers (61.4 per cent) were still repaying their redemption loans[2].

2 The dossiers concerning the Wierthe and Vechelde sugar plants and their shareholders are preserved in the States Archive of Wolfenbüttel. Concerning the shareholders, dossiers from groups 46 Neu (Hypothekenbücher [predecessors of the land registers]) and group 47 Neu (Amtshandelsbücher [books where the official duties of the local magistrates were notified]) have been used. Particular use has also been made of Files 40 Neu 21, Nr. 10129 and 10133 concerning the Vechelde and Wierthe sugar plants.

The ability of the Brunswick peasants to honour their financial commitments, originating in the redemptions, and to participate in setting up sugar plants resulted from the special nature of the redemptions and the redemption loans. First of all, in the duchy of Brunswick relatively low rates were set for evaluating former taxes and payments in kind. Furthermore, the farmers got a 12 per cent discount on the annual amount of their payments in kind because the lords received only the equivalent of their net revenue. Moreover, when the redemptions were calculated, it was estimated that the interest rates for redemption loans would come to 4 per cent; in fact interest rates had dropped to 3.5 per cent by 1839. Finally, the Brunswick state offered the peasants special redemption loans. These loans had a characteristic that distinguished them from redemption loans offered by other German countries: they were not repaid with a fixed annuity consisting of a rising rate of repayment and a decreasing rate of interest, but by a fixed rate of repayment and a decreasing rate of interest. Therefore the burden of honouring the financial commitments decreased little by little.

As a result of the low cost for the former payments in kind, in combination with the discount applying to them and the characteristics of the redemption loans, the financial burden of the farmers rose by only 3.05 per cent in the first year after redeeming the annual dues. In the case of the taxes, the burden rose by 13.55 per cent in kind, which meant that the annual burden of the peasants decreased within five years, to less than the value it had been before the redeeming took place, so that it took only fifteen years to repay the taxes. The weight of repayment had also been lightened by the good agricultural economic situation and rising prices for farm products between 1840 and 1870. Consequently, when the sugar plants were founded the burden of the redemptions was much lower than the former dues.

Table 6.1. Proportion of peasant shareholders still repaying redemption loans. The Vechelde and Wierthe sugar plants

Still repaying loans	61.4%
Repayment already finished	38.6%

Source. States Archive of Wolfenbüttel. See footnote 2..

Even more impressive is the amazingly small number of peasants who needed additional loans in order to buy shares in the sugar plants. Out of the number of 62 farmers still repaying their redemption loans, only 12 (19.4 per cent) required additional loans. More than 80 per cent of the peasants still repaying their redemption loans were able to finance the specialisation in sugar-beet production with their own money; 5 (12.8 per cent) of those peasants who had already finished repaying their redemption loans used additional loans to finance share purchases in those two sugar plants.

Table 6.2. Proportion of peasants shareholders needing additional loans and still repaying redemption loans. The Vechelde and Wierthe sugar plants

Additional loans	19.4%
No additional loans	80.6%

Source. States Archive of Wolfenbüttel. See footnote 2.

There are two possible explanations why Brunswick peasants could participate in the specialisation of farming even while they were still repaying their redemption loans. Firstly, the encumbrance from the repayment of the redemption loans was only slightly higher than one of the former dues (dues in natural produce 3.05 per cent; dues paid in money equivalents 13.55 per cent (Kulhawy, 2007: 123). Secondly, the burden from the repayment of the redemption loans shrank yearly for most of the Brunswick farmers who were the debtors of the state-owned Brunswick Mont-de-piété (Leihhaus). In contrast to the state-owned banks of other German states that financed redemptions, the Leihhaus did not demand fixed annuities until the repayment ended. Instead, it demanded a fixed annual amount of repayment and in order to make this possible, the interest charge fell each year. So the burden originating from the former dues fell below its original amount after five years (for dues once paid in natural produce) or after sixteen years (for dues paid in money equivalents). The favourable economic conditions enjoyed by German agriculture in the middle of the nineteenth century and falling interest rates made the repayment even more bearable. Generally speaking, the conditions of the Leihhaus redemption loans seem to have been so favourable to the farmers from all parts of the duchy that many of them preferred to invest in their estates instead of paying off these loans.

The high degree of agricultural specialisation in the duchy of Brunswick is shown by the development of the acreage cultivated with sugar beet. Starting near zero at the end of the 1840s this figure rose to 10,000 hectares in 1880 and to 26,000 hectares by 1910 (Pommer, 1912: 340). The share of the duchy's acreage cultivated with sugar beet rose significantly, from 6.9 per cent in 1878 to about 14.6 per cent in 1911 (Buerstenbinder, 1881: 216; Pommer, 1912: 339). These figures differed from other parts of Germany where the share of the acreage cultivated with sugar beet rose from approximately zero in the middle of the nineteenth century to 2.2 per cent in 1913.

The figures for certain regions of the duchy are even more impressive. Early centres of sugar production were the judicial districts of Schöningen, Schöppenstedt and Wolfenbüttel. In 1880, the Schöningen area already had 20 per cent of the acreage undersugar beet (Buerstenbinder, 1881: 216). In the Schöppenstedt area it amounted to 14 per cent and in the Wolfenbüttel district it came to 13.6 per cent. In 1910 in the Wolfenbüttel administrative district 24.7 per cent of the acreage was cultivated with sugar beets (Achilles, 1993: 218). This figure was the highest of all German rural districts. In the administrative district of Helmstedt, which was in fifth place for sugar cultivation at that time, sugar beet cultivation came to 16.2 per cent.

In the duchy of Brunswick, sugar beet growing was usually combined with more intense cattle and sheep breeding. Both were fed with the left-overs of sugar production. The young calves were bought in eastern Prussia and raised in the barns of farmers who cultivated sugar beet. Later, the cattle – as well as the sheep – were sold in Brunswick or in the Rhine and Ruhr area.

Another important specialisation in the agriculture of the duchy of Brunswick, principally in the administrative districts of Brunswick and Wolfenbüttel, which were the most fertile areas of the country, was the development of small plants for preserving food production, which stimulated vegetable-growing and cattle breeding (Kanzow, 1928: 69, ff.; Pommer, 1912, 342). Between 1851 and 1880, in Brunswick, 14 firms of this kind were set up to turn asparagus, peas, beans, mushrooms, cabbage, turnips and cattle into preserved food. By 1928, the number of companies had risen to 43 in Brunswick and another 38 in the remaining parts of the duchy. As a result, by 1910 the share of total acreage under vegetables in the duchy had reached 4.8 per cent. As mentioned above, the growing area was concentrated in the Brunswick and Wolfenbüttel area, which means that the percentage there was much higher.

Agricultural development was so strong that it stimulated the industrial development of the duchy (Kanzow, 1928: 11; Schildt, 2000: 798). Firstly, it provided the incentive to set up a food-processing industry; secondly, the demand for combustibles from the sugar processing plants encouraged the development of the Helmstedt brown-coal mining area; thirdly, demand from the agricultural sector promoted the development of factories specialising in agricultural machines and agricultural consumer goods.

VII. Conclusion

In conclusion, we can say that until the middle of the twentieth century real specialisation of agriculture in the duchy of Brunswick took place only in its fertile areas. In the eighteenth century government attempts to encourage modernisation and specialisation were not successful, as the social and economic conditions of the farmers had not changed. Knowledge about agriculture and about how to organise agricultural reforms was limited. Last but not least, the ducal administration was not yet able to carry out reforms of this sort.

Significant change came in the 1830s. Having learnt from other German states and other European countries, the ducal administration succeeded in implementing agricultural reforms. These reforms were supported by new transportation networks, new techniques of agricultural education and a favourable economic climate. Consequently a specialisation combining the cultivation of sugar beet with cattle breeding and vegetable production got its start in the fertile regions of the duchy.

It is hard to weigh the relevance of endogenous versus exogenous factors in this process of specialisation. On the one hand factors like the quite large size and good equipment of farms, the fertile soil, the favourable climate and the peasants' experience in intensive cultivation techniques could be cited to support the importance of endogenous causes. Moreover, the enormous expansion of the transport network organised by the villages and local administrations was considerable proof of the active integration of peasants into the market process. On the other hand, the work of the agricultural club; the early example of the neighbouring Prussian province of Saxony, where sugar beet had been successfully cultivated; the early development of a state-owned railway system, assisting market integration; and the agricultural reforms, including the obligation in the redemptions of honouring financial commitments, are strong arguments for the importance of exogenous factors. Without the incentives of the market the specialisation would not have taken place. But likewise it would not have taken place without the fertile soil, the ideal climate and the farmers' experience in intensive cultivation methods. It therefore, be seems appropriate to say that both endogenous and exogenous factors played an important role in the process of specialisation.

This agricultural development had a great influence on the further industrial development of the duchy, as it encouraged the setting up of sugar factories, food processing and preserving plants and factories specialising in agricultural machines and the production of agricultural consumer goods. This form of economic development was typical for regions where sugar beet was cultivated, as can be seen in the neighbouring Prussian province of Saxony and, to some extent, in the Spanish region of Zaragoza (Müller, 1979: 24; Pinilla, 1995: 386).

Bibliography

ABEL, Susanne (1990), 'Zum Wandel der Landwirtschaft im Herzogtum Braunschweig im 19. und 20. Jahrhundert', *Südniedersachsen, Zeitschrift für Kultur un Heimatpflege,* 18, p. 131-141.

ABEL, Wilhelm (1978), *Agrarkrisen und Agrarkonjunktur. Eine Geschichte der Land- und Ernährungswirtschaft Mitteleuropas seit dem hohen Mittelalter,* [Berlin, P. Parey, 1966], 301 p.

ACHILLES, Walter (1993), *Deutsche Agrargeschichte im Zeitalter der Reformen und der Industrialisierung,* Stuttgart, E. Ulmer, 397 p.

ACHILLES, Walter (1978), 'Die niedersächsische Landwirtschaft im Zeitalter der Industrialisierung 1820-1914', *Jahrbuch,* 49, p. 7-26.

ALBRECHT, Peter (1980), *Die Förderung des Landesausbaues im Herzogtum Braunschweig-Wolfenbüttel im Spiegel der Verwaltungsakten des 18. Jahrhunderts (1671-1806),* Braunschweig, Waisenhaus-Buchdruckerei und Verlag, XIX-669 p.

ALBRECHT, Peter (2000), 'Das Zeitalter des aufgeklärten Absolutismus (1735-1806)', in Horst-Rüdiger JARCK, Gerhard SCHILDT, (eds), *Die Braunschweigische Landesgeschichte. Jahrtausendrückblick einer Region*, Braunschweig, Appelhans, p. 575-610.

BRAKENSIEK, Stefan (1991), *Agrarreform und ländliche Gesellschaft. Die Privatisierung der Marken in Nordwestdeutschland 1750-1850,* Paderborn, PF. Schöningh, IX, 516 p.

BUERSTENBINDER, Richard (1881), *Die Landwirtschaft des Herzogtums Braunschweig,* Braunschweig, VII-354 p.

BUERSTENBINDER, Richard (1883), 'Über die gegenwärtigen bäuerlichen Verhältnisse im Herzogthum Braunschweig', in *Bäuerliche Zustände in Deutschland,* Leipzig, Verein für Socialpolitik, 2, p. 79-105.

ELLERBROCK, Karl-Peter (1993), *Geschichte der deutschen Nahrungs- und Genussmittelindustrie 1750-1914,* Stuttgart, Franz Steiner, 477 p.

JIMENEZ-BLANCO, Ignacio (1986), 'La remolacha y los problemas de la industria azucarera en España, *1880-1914', Historia Agraria de la España Contemporanea*, Madrid, Ministerio de Agricultura, III, p. 280-316.

KANZOW, Georg (1928), *Grundzüge der braunschweigischen Industrie. Ein Beitrag zur Wirtschaftskunde Niedersachsens,* Hannover, Selbstvverlag, 99 p.

KRAWINKEL, Max-Ferdinand (1994), *Die Rübenzuckerwirtschaft im 19. Jahrhundert in Deutschland,* Köln, Botermann and Botermann, XLII, 307 p.

KULHAWY, Andreas (2007), ''Bauernbefreiung' auf Kredit: Aus der Praxis des Braunschweigischen Leihhauses 1834-1930', in Jürgen SCHLUMBOHM (ed.), *Soziale Praxis des Kredits, 16.-20. Jahrhundert*, Hannover, Hahnsche Buchhandlung, p. 121-151.

MARTIN-RODRIGUEZ, Manuel (1982), *Azúcar y descolonización. Origen y desenlace de una crisis agraria en la vega de Granada,* Unpublished Ph. D. Granada, 390 p.

MEYBEYER, Wolfgang (2000), 'Die Landesnatur. Territorium – Lage – Grenzen', in Horst-Rüdiger JARCK und Gerhard SCHILDT (eds), *Die Braunschweigische Landesgeschichte. Jahrtausendrückblick einer Region*, Braunschweig, Waisenhaus-Buchdruckerei und Verlag, p. 21-44.

MÜLLER, Hans-Heinrich (1979), 'Zur Geschichte und Bedeutung der Rübenzuckerindustrie in der Provinz Sachsen im 19. Jahrhundert unter besonderer Berücksichtigung der Magdeburger Börde', in Hans-Jürgen RACH, Bernhard WEISSEL (eds), *Landwirtschaft und Kapitalismus; Zur Entwicklung der ökonomischen und sozialen Verhältnisse in der Magdeburger Börde vom Ausgang des 18. Jahrhunderts bis zum Ende des Ersten Weltkriegs*, Berlin, Halbband, 2, p. 9-61.

MÜLLER, Uwe (2000), *Infrastrukturpolitik in der Industrialisierung. Der Chausseebau in der preußischen Provinz Sachsen und dem Herzogtum Braunschweig vom Ende des 18. Jahrhunderts bis in die siebziger Jahre des 19. Jahrhunderts,* Berlin, Duncker & Humblot, 585 p.

PINILLA, Vicente (1995), *Entre la inercia y el cambio. El sector agrario aragones, 1850-1935*, Madrid, Ministerio de agricultura, pesca y alimentación, Secretaría general técnica, Centro de publicaciones, 546 p.

POMMER, Emil (1912), 'Der Stand der Landwirtschaft im Herzogtum Braunschweig', *Jahrbuch der Deutschen Landwirtschafts-Gesellschaft,* 27, p. 337-354.

SCHILDT, Gerhard (2000), 'Die Industrialisierung', in Horst-Rüdiger JARCK, Gerhard SCHILDT (eds), *Die Braunschweigische Landesgeschichte. Jahrtausendrückblick einer Region*, Braunschweig, Waisenhaus-Buchdruckerei und Verlag, p. 787-821.

SCHNEIDER, Karl Heinz, SEEDORF, Hans Heinrich (1989), *Bauernbefreiung und Agrarreform in Niedersachsen*, Hannover, Niedersächsische Landeszentrale für politische Bildung, 124 p.

SIEBENBRODT, Willi (1938), *Die Braunschweigische Staatseisenbahn. Zur Jahrhundertfeier der ersten deutschen Staatsbahn Braunschweig–Wolfenbüttel,* Hannover, herausgegeben von der Reichsbahndirektion, 112 p.

VENTURINI, Carl Heinrich Georg (1847), *Das Herzogtum Braunschweig in seiner gegenwärtigen Beschaffenheit,* Helmstedt, C.G. Fleckeisensche Buchhandlung, VIII-325 p.

WITTICH, Werner (1896), *Die Grundherrschaft in Nordwestdeutschland,* Leipzig, Duncker & Humblot, 461-143 p.

7. Intensification without specialisation. The development of a South-Bohemian manorial estate in the nineteenth century

Pavel Matlas

I. Introduction[1]

The nineteenth century was a time of revolutionary development in Czech agriculture, caused not only by scientific and technical progress, but also by the political events of the revolutionary year 1848, which led to considerable changes in legal and social conditions in the Czech countryside.

Financial compensation paid to estate owners for the loss of feudal privileges was meant to facilitate the reorganization of agricultural production, which had hitherto depended on a work force of serfs performing manorial labour. How was the compensation spent, and what form did these changes take in the management and structure of the manorial estate? What difference did they make to conditions on the manorial estate, the basic organizational unit of agricultural production? To answer these questions, we have chosen as a model a typical South Bohemian manorial estate. The owners of the estate, the Dukes of Schwarzenberg, were among the largest land owners in the Habsburg monarchy. Their surviving and extensive archives made it possible to trace the changes in agricultural production on selected manorial estates from a micro-historical perspective. Despite this work it was not possible to prove that there was specialisation of agricultural production in this area. But it is clear that, with the abolition of serfdom, which had presented an obstacle to the use of human labour, the laws applying to labour, and the condition of workers improved. This is why this process should be called intensification and not specialisation.

At the same time, this contribution has been conceived as a case study, in the context of the present state of research into these problems. Further work needs to

1 This paper is an amended and enlarged version of a paper first published in *Prager Wirtschafts- und sozialhistorische Mitteilungen*, 15, 2012, p. 28-40.

Agricultural specialisation and rural patterns of development, ed. by Annie Antoine, Turnhout, 2016 (*Rural History in Europe*, 12), p. 135-149.

BREPOLS PUBLISHERS DOI 10.1484/M.RURHE-EB.5.112265

be done into the agricultural history of Central and Eastern Europe, especially for the territory of the former Polish-Lithuanian State, where manorial demesnes based on corvée labor were also established during the time of what is termed the 'second serfdom' (Mencel, 1987: 53-77). Such research, if it leads to similar observations, is the only way to make our results useful for the purposes of comparison and for valid overall general conclusions[2]. However this task is outside the range of the present contribution and beyond the scope of current knowledge.

II. General political, social and legal background

In this area, unlike most parts of Eastern Europe, feudal relations were not influenced by the general loosening of feudal bonds in the regions east of the Elbe. On the contrary, they developed in a quite different direction. The course of the Thirty Years War was accompanied by a profound economic decline, together with the formation of servile estates, based on increased feudal rents. Undoubtedly, serfs saw manorial labour, or tasks which had to be worked off on the manorial land, as the most oppressive burden of all. Manorial lords easily obtained a cheap work force, but also saved money since all the cost of agricultural implements was covered by the serfs, who were forced to use their own teams of draught animals. Manorial lords also profited by the fact that their land, termed demesne land (land cultivated by the lords), was not subject to ordinary taxation, unlike what was called rustic soil, tilled by serfs (Kočí, 1957: 63).

Serfdom finally ended in 1848. According to an article in the constitution of 25 April of that year, the issue of serfdom was to be settled by the newly formed imperial court. The court was dominated by liberal politicians, who maintained that private ownership was indisputable. The law of 7 September 1848 did away with serfdom and all the power the aristocracy had over their subjects; it also eliminated differences between demesne and serf lands. Needless to say, the law required aristocratic landowners to be paid financial compensation in return for the abolished obligations, services, instalments and payments in kind, which had hitherto been the serfs' responsibility. The amount of this compensation was fixed by regulations published in 1849 (Roubík, 1959: 161; Kazbunda, 1929: 43-58).

---

2 The best summary of older historiography on this topic for the territory of the Polish-Lithuanian state is A. Maczak (1981), (ed.), especially the sections 'Mechanizacja rolnictwa/mechanisation of Peasantry' by Julian Bartyś, p. 505-507 and 'Rolnicztwo/Peasantry', by Janina Leskiewiczowa, p. 185-208, 213). In the early 1990s historical writing in the countries of the former Eastern bloc began to diverge from certain pivotal themes of the official Marxist historiography. The agricultural history of the countryside, whose interpretation had been oriented towards class antagonism, represented one of these topics. For that reason, no attention was paid to this topic in specialized papers for at least two decades.

After the amount of compensation for serfdom had been set, it was divided into three equal parts. One-third was given to aristocratic landowners, as a lump-sum compensation for the services they had provided for their serfs (charities, helping the poor and sick, advances for sowing, management of custodial and guardian issues). The government used special funds to pay off the second third on the farmers' behalf in quarterly instalments scheduled over 40 years. New landowners were obliged to pay off the final third of the compensation themselves (Krofta, 1949: 428 et seq.). Nonetheless, they were still able to postpone the payments. Because of social tension in the countryside, the amount had to be split up into long-term payments spread over twenty years. The average annual financial charge therefore varied between a reasonable 2 and an extravagant 10 gulden per year (Kalousek, Archiv český, 1910: 720 and following).

The enforcement of New Year Patents of 31 December 1851, or rather, their implementation, eventually came to be the only result of the revolutionary period of 1848-1849 but it was a crucial and lasting one. The proclamation of these letters patent brought an end to all attempts at reform, and established a new version of royal absolutism (Urban, 2003: 225-228). The abolition of servile dues and patrimonial ties had irreversible economic and social consequences. The way it was enforced went in the direction of the 'Prussian way', that is, the transformation of a feudal estate into a capitalist estate with some remaining feudal features (Jakubec and Jindra, 2006: 102). The development of small agricultural businesses was hindered by the duty of former serfs to pay compensation, whereas former aristocratic landowners were left with considerable financial means. These financial means resulted in a rapid transition to capitalism. The largest landowners in Bohemia – the Schwarzenbergs – were awarded compensation amounting to 2.25 million gulden[3].

III. The outline of historical development and natural conditions

Two of the reasons why there was no specialisation of agricultural production in this area were climate and soil quality. South Bohemian latifundia – in their size and the way they were organized – constitute a distinctive phenomenon in the agrarian history of Central Europe. Their history dates from the period of medieval colonization of the area; further concentration of land possession took place towards the end of the

[3] The total financial compensation in Bohemia came to about 90 million guldens, 80 per cent of which went to former aristocratic landowners, the rest being divided among various public institutions. The fact that 578,341 individuals paid compensation and 22,762 persons and corporations, including 1231 overlords claimed compensation rights, demonstrates the considerable scope of the whole enterprise. URBAN (2003: 144).

Middle Ages during the Hussite Revolt[4]. When the Dukes of Schwarzenberg became the new owners of the Třeboň estate, they further extended their holdings.

The estate reached its greatest size in the 1780s. In the year 1787, the government transferred to Jan Nepomuk Schwarzenberg the possession of the former Augustinian monastery in Třeboň, which had been closed down by Emperor Joseph II two years before (Bílek, 1893: 157-158; Kadlec, 2004: 16-22). This acquisition was the last change in the territorial extent of the Třeboň estate. So, at the end of the feudal period, the demesne comprised 92 villages, 3 hamlets, 4 towns and spread out over roughly 1,700 square kilometres (rustic land together with demesne) (Kraft, 1872: 191).

Table 7.1. The extent of agricultural land of the Třeboň manorial estate, 1840

Fields	1,719.0 ha
Meadows	9,443.0 ha
Grasslands	852.5 ha

Source. SOMMER (1841: 65-66).

Their Třeboň manorial estate had thus gradually spread over the whole of the Třeboň basin. However, the soil composition of the basin, together with the landscape relief and climate, do not offer ideal conditions for agricultural production (Jakubec and Jindra, 2006: 102). In the south, near the town of Soběslav, there are extensive peat deposits, which only approximate the quality of poor swamp humus to the north, towards the town of Třeboň, the centre of the demesne. Only in the central part of the Třeboň basin, do we find more diversified soil, with heavy clay-type layers alternating with limonite and aluminium ferric ores. Flat granite islands crop up only along the western limits. Clay type soil is found to the north, but the topsoil is made of gravel layers or, much less frequently, clay-type layers. As a result, the arable land cannot absorb humidity and water stands for a long time (Kraft, 1872, 194; Heske, 1909: 10-12). Moreover, as far back as we can trace in the records, the Třeboň basin has always suffered periodic inundations. This originally marshy land was frequently flooded by the Lužnice river, fed from the Novohradské Hory mountains, which are subject to heavy rainfalls. But with the formation of a unique pond system and the drainage of extensive areas agricultural land was greatly extended (Šusta, 1995: 212; Hule, 2000: 205). Towards the end of the feudal period, the Třeboň landlords administered as many as 4,000 hectares. This was so-called demesne soil, which was worked in

[4] The Třeboň estate covered 43 residential localities in the latter half of the fifteenth century. Later on, in the early 1520s, after the purchase of a few small farms, it came to include as many as 57 localities. ŠIMUNEK (2005: 100-101).

15 demesne farms, with an average surface of 262 hectares[5]; the whole extent of agricultural land of the Třeboň manorial estate ran to more than 12,000 hectares.

IV. Sowing procedures

As a result, development went in a different direction from the path taken by grain production in the fertile areas of Polabí, hop growing in the region of Žatecko, or wine growing in south Moravia (Albert, 1970: 47-68). Higher production in the region under study was reached thanks to more advanced agro-technical methods and widespread mechanization in particular (Pátek, 1971: 51-76).

The three-field rotation system had been the backbone of the original production in force since early times. Based on cyclic alternations of spring corns, winter wheat, and fallow, it could not be made to yield high production. The first half of the nineteenth century saw the introduction of a new polyvalent field system, which included the replacement of at least part of the fallow (the so-called black or pure fallow), with fodder plants or root-crops (green fallow). None the less, agricultural production was still based on the three-field system, though refined to some extent. It was only the introduction of the alternate field system that brought a fundamental change in the efficiency of agricultural production. The first mention of its introduction into Bohemia dates from the year 1808 in Libochovice. The continual and non-violent transformation from the three-field system into alternate farming was facilitated by the rise in agricultural prices, caused by the Napoleonic wars (Jakubec and Jindra, 2006: 96-97, 104; Horsky, 1861: 64). Its significance lay mostly in the fact that for every subsequent crop its most suitable antecedent was selected; moreover, the same crops were not sown in succession. The Schwarzenbergs were the first to benefit from the alternate sowing system. From 1836, when František Horský the manager of Libějovice estate, began to put the principles of the technique into practice, to 1844, estate output rose by almost 100 per cent (Horsky, 1847: 19). By then there was no obstacle to introducing alternate farming in the other Schwarzenbergs' estates. Coincidentally, Třeboň began to employ this technique in 1848, the year of the Revolution. Towards the end of the 1860s, two types of the ten-phase crop plan were carried out in the region, depending on soil quality.

The local soil was particularly suitable for growing root-crops, the most important of which were potatoes and turnips. Turnips, notably the so-called cow's turnip, were grown for feeding animals. However, the more demanding sugar beet was grown as well (Kutnar, 2005: 22-32). With the growing emphasis on protection against

[5] State regional archives of Třeboň: Estate funds [henceforth Ef], series IC 6M β 3a., Summary report of the manorial estate economy for the year 1840 (25 January 1842).

pests, papilionaceous fodder plants appeared in sowings during the first half of the nineteenth century (Křivka, 1989: 82-406). Before clover came in, first sown in the region of Třeboň during the 1820's, lucerne was known as a cultivated crop in Central Europe. Typically, legume plants would appear in the Třeboň sowing system. These plants, rich in albumen, were used in the animal production as winter fodder (Loudil, 1980: 20-41).

Table 7.2. Two patterns of the ten-side sowing plan

	A.	*B.*
1	green fallow	green maize
2	beet (sugar)	winter corn
3	winter corn	clovers
4	clovers (red and 'Luzern')	mixture of clover and grasses (Thymoteus and Ray)
5	clovers (Schnitt)	grasslands
6	winter corn	
7	potatoes, stock beet, cabbage, broad bean	
8	barley, summer rye, oat	
9	vetch, sweet pea, buckwheat	
10	oat	

Source. Kraft (1872: 197).

Table 7.3. The plant representation in the long-term average in the middle of the nineteenth century

	percentage
oat	27.1
rye	22.6
barley	12.8
wheat	10.5
root crops	15.3
fodder crops	8.5

Source. Hanuss (1875: 48-19)[6].

6 See closely SOA Třeboň, Hanus's inspection fund, inv. no. 28.

As a matter of fact, the rotation mentioned above included a combination of all basic species of agricultural plants. At the same time, however, the alternate system still contained the three-field system as parts of the whole agricultural area were used as fallow, meadows, and pastures. Potential specialisation of agricultural production was hindered by lower soil quality and unfavourable climate. Certain crop preferences and concentration on certain forms of agricultural production would have over-worked the soil. As a result, all species of agricultural plants had to be kept in balance within the limits of alternate farming. The feudal tradition of the demesne, which had always been a closed, economically self-sufficient unit, seems to have acted as yet another obstacle for agricultural production to specialize. Since the middle of the nineteenth century, the Schwarzenberg administration therefore attempted to increase the demand for agricultural products in various ways. Instead of specialisation, the Schwarzenbergs chose to intensify production.

V. Draining

The money obtained as compensation for the loss of feudal privileges, was used mainly to pay for extensive drainage systems. As early as 1850, Johann Spies, a builder who worked for the Dukes, assembled an improved machine for pressing drainpipes, adapted from an English model. The following year, 26 kilometres of pipes drained 93.5 hectares of agricultural land, where extreme humidity had previously made farming unprofitable. The output of the first crops put a clear stop to any doubt about recovering the expenditure. The experience with the ample autumn rains of 1851 showed that water left these areas three days faster than it left the areas not drained hitherto. Above all, the drained area allowed deeper ploughing and cultivation of the arable land; fewer weeds grew, seedlings sprouted up much faster, and there was more time for plants to ripen and therefore yield more produce[7]. It is no surprise that the Duke of Schwarzenberg ordered Spies to assemble a similar machine for another Schwarzenberg estate in Netolice. By 1856, at which point further work had to be halted for a period of 12 years, 184 kilometres of pipes drained 337 hectares of agricultural land. Reportedly, the Třeboň region was the first on the Continent to have used the new drainage system. By the end of the nineteenth century, 523 kilometres of drainpipes had been laid in the region of Třeboň (Kalný, 2001: 131-132)[8].

It is no coincidence that the beginnings of the English drainage method in the Třeboň region date from the years immediately following the abolition of serfdom. The drained fields now produced a higher annual yield, making up for the losses in

[7] For a detailed evaluation of particular plants, crops and the impact on agricultural methods, see farming reviews, 31 December 1852 and 31 January 1857. Ef Třeboň, IC 6M beta 3b/2.

[8] SOA Třeboň, Hanus's inspection fund, inv. no. 36.

vegetable production, which was significantly jeopardized by the sudden decline of the work force, comprising a huge number of serfs. Many of them later worked as hired labourers, draining agricultural land, a process partly financed by compensation paid for the abolished labour dues (Matlas, 2006: 250-253).

VI. Mechanization, ploughing methods

Agricultural mechanization was another sphere into which Czech landlords invested considerable capital, money which came from compensation for lost feudal privileges (Pátek, 1971: 59-68). Until 1848, farming was entirely dependent on the serfs' traditionally simple equipment, with the exception of ploughing. Immediately after the swing-plough was invented by the Veverka brothers in 1827, it became an indispensable part of the equipment on the Schwarzenberg estates. The agricultural inventory of the Lhota estate (one of the 16 manorial estates) mentions two tracers as early as in 1829[9].

Agricultural production was also increased by improving the procedures of farming technology. The quality of the future crops depended especially on suitable technique and the correct timing of arable work. At the beginning of the 1840s, Franz Horský, Schwarzenberg's agricultural inspector, experimented with double ploughing, performed by two ploughs in line. However, ploughing shallow arable in this way brought to the surface the infertile layer from the subsoil. To cultivate arable of this kind, a large amount of manure was needed. Consequently, Horský discarded this method, and recommended a simple deep ploughing, the result of which was to bring the most fertile layer of earth to the surface. This, as Horský rightly assumed, was to be found in the topsoil immediately under recently ploughed earth. (Horsky, 1860: 69-71). As plants could reach nutrients much earlier, the crops increased by up to 63 per cent in five years. According to Horský, the deep ploughing method was advantageous because it enabled the land to resist weeds; the soil was loosened deeper, which helped the root system penetrate better, while it also protected plants against effects of dry spells, or, vice versa, against unwanted dampness (Horsky, 1860: 71).

From 1853, František Horský introduced a new system of ploughing on all the estates under his management. This new system was a combination of shallow and deep ploughing in two successive layers of earth. For that purpose, he suggested that two smallish narrow adjustable duck-foot shares should be inserted under the main share of the swing-plough. When using a swing-plough adjusted this way, the farmer ploughed in manure or stubble to a depth of only 2 inches, while at the same

9 Agricultural inventory of Lhota estate on Třeboň manorial estate, 1 July 1829. Ef Třeboň, ID 6I 3.

time cultivating the soil to a depth of 12 to 14 inches (Horsky, 1860: 75-76). As a result, manure would stay in the top layers of arable, where it was closer to the air, and decomposed faster; humus, the outcome of the whole process, was thus easily accessible to the seedlings (Horsky, 1860: 73; Horsky, 1862: 55-59).

VII. Sowing work and harvesting

Horský's discoveries and technical inventions influenced sowing procedures in a similar way. As early as the first half of the nineteenth century, while in Schwarzenbergs' service, he became familiar with the workings of grain seeders[10]. He designed a number of substantial improvements to the Garret seeder, at that time the best known and most widely used seeder on Czech estates. As a result of these improvements the machine weighed less, thereby making smaller demands on draught power, and cost less to buy. The machines were more adapted to cultivation work, and their greater flexibility enabled them to adjust to both the character of the soil and the kind of seed used. (Horsky, 1860: 246-249). As machines of this kind could insert grain into the soil evenly and at a suitable depth, crops appear to have increased by 15% (Vilíkovský, 1936: 27)

The slowest development took place in the farming technology used in harvesting. In southern Bohemia, down into the nineteenth century, grain was reaped by the grass scythe, into the handle of which was interwoven a hazel twig. In this way stalks could be handled with care, preventing grain from falling out of the ears of grain. It was only with the introduction of reaping machines that harvesting became more efficient. As early as 1852, Antonín Burger from Vienna, a merchandiser and exclusive agent of the American McCormick company, offered reaping machines to the Schwarzenbergs' estates. With only a pair of horses and two people working the machine, as many as 20 acres (or 14 upper-Austrian acres of grain a day)[11] could be mowed. The prospectus assured the owners of former estates, who now had to do without their serfs' statutory labour, a significant reduction in their work force. A machine cost only 400 Guldens of conventional currency, and Burger's company offered a bulk discount of 50 Guldens apiece[12]. Burger found the same strategy successful in promoting all his products, as they ensured reduced labour costs[13]. In the following years, the Třeboň estate bought a great number of new cultivating tools. Besides the above-mentioned seeders and ploughs, these included various devices such as soil-looseners, coulters, potato diggers, pickers, weeders, rollers, manure spreaders, dung-water pumps, sprayers, and the like.

---

10 For their use on Schwarzenberg estates, see receipts for their maintenance, e.g. Ef Třeboň, ID 6I 2b.

11 Approximately 8.057 hectares.

12 Ef Třeboň, ID 6I 2b.

13 *Ibid.*

Subsequent processing procedures did not lack improvements either. In the 1850's the one-hundred-year-old flail threshing machines began to be rejected in favour of mechanized threshing machines. The most widespread were the relatively cheap Hensman manual whim, or flail, and Barrett's all-metal threshing machines, which were produced in Prague in the latter half of the nineteenth century (Tempír, 1986: 55-61). However, more efficient crush threshing machines were preferred to them on the extensive South Bohemian estates. The Třeboň estate used one such threshing machine, which was driven by two pairs of horses and attended by six hands. The machine could thresh 78 shocks of oats or 66 shocks of barley in two and a half days[14].

VIII. The introduction of the steam plough

Nonetheless, the real breakthrough in the mechanization of agricultural production occurred with the steam engine. Reportedly, the first steam engines turned up on some Schwarzenberg estates at the end of the 1840s, but it was not until one or two decades later, when they became capable of high performance, that they came into their own[15]. One of the points in favour of the steam engine was that it could drive other machinery, such as belt conveyers, winnowers, pumps, and the like. The steam plough stood at the pinnacle of all technical equipment at that time. It was formed by a system of two mobile locomotives. Unlike true locomotives however, loco-mobiles, as the word implies, were not capable of independent movement; as a matter of fact, they were simple steam engines placed on a wheeled undercarriage. These portable steam engines were drawn into place on the edges of a field by a two-horse team. Using ropes, they drew a double-arm plough with three or more shares across the field. The whole unit had to be attended by at least three workmen; two of them to run the steam engines, and the third to watch the ploughing implement.

Schwarzenbergs' central administration began to show great interest in the latest inventions of agricultural technology in the first half of the 1870s. In 1872 the Stone & Lythall Company of Prague (Smíchov) became sole agent of the English company Ransomes, Sims and Head, and offered the steam plough to the Schwarzenberg manorial estates. The machine was capable of ploughing ten acres of land a day. However the sale did not take place, because the purchase price was extremely high, almost 29,000 gulden, and the cost of overhead was equal to that of a team of oxen[16]. But when the steam engine was finally introduced on one of the estates in the fiscal year 1875/76 the initial experience was indeed impressive. Running the English

[14] Report, 15 February 1852. Ef Třeboň, ID 6I 2b.

[15] *Ibid.*

[16] SOA Třeboň, Hanus's inspection fund, inv. no. 37.

Howard-built steam plough for one day amounted to about 45 per cent of the cost of operating two teams of oxen. Accordingly, it was estimated that its purchase price, 5,200 gulden, could be recovered in a mere 5 years (Matlas, 2006: 255-260).

Moreover, the machine would work longer than oxen, which needed a three-hour break at noon to rest and graze. A team of oxen lost 10 kilos of its live weight during one day of ploughing. Such wastage meant a considerable loss, since even draught oxen were sold as butcher's meat after reaching six years of age. Another important advantage of the steam engine was the fact that the amount of ox manure increased considerably. Not only did un-harnessed Spanish breed oxen retain their weight, they also produced, with the same fodder, 110 kg (269 per cent) more manure in ten days (Hanuss, 1877: 23-31). The attention paid to this problem reflects the importance of manure in agriculture at that time.

IX. Fertilization

The increased demand for natural fertilizers was a result of the alternate farming system. The land, now cultivated more intensively, and no longer lying fallow every third year, would have lost its productivity without supplementary nutrients. Even though the number of cattle increased, it could no longer meet the greater demand for fertilizers (Hanuss, 1860: 105-109, 121-126; Kraft, 1872: 202).

Table 7.4. The cattle in 1900

	Number	*Percentage*
Cattle	33,486	74%
Small cattle	8,123	18%
Mutton	2,602	6%
Horses	1,052	2%

Source. HESKE (1909: 23-24).

Having heard encouraging news from Belgium, the Schwarzenbergs arranged for a bone grinder-crusher to be installed in the Třeboň region in 1830; their aim being to produce bone meal for fertilizer. However the use of bone meal did not become significant in the end, since mineral fertilizers yielded better results. Their efficiency was proved in experiments carried out ever since 1789, first with plaster on the North Bohemian estate of Postoloprty, and later on the South Bohemian estate of Český Krumlov, where the effect of fertilization with lignite was tested in 1818 (Vilíkovský, 1936: 21).

The introduction of artificial fertilizers meant a significant difference for the Třeboň region, whose soil could not boast any calcium. As guanine, which had been imported since 1843, proved successful, Adolf Schwarzenberg went on to produce 'pudreta', a powder fertilizer, which finally took the place of expensive imported fertilizers on the estates[17]. In 1857, the Třeboň manorial estate participated in comparative fertilization experiments with Chile saltpetre, pudreta, bone meal, and oil-cake meal. Potassium salts proved to be partly useful as did as kainite (a kind of potassium salt). Nevertheless, superphosphates were far more important for agriculture; initial industrial production of superphosphates dates from the 1870s (Jakubec and Jindra, 2006: 138).

X. Conclusion

Developments on the Třeboň estate show that the Schwarzenbergs' manorial estate was one of the most advanced establishments in Bohemia. However, increased agricultural output was not achieved by the specialisation of agricultural production, which was impeded by unfavourable climatic and soil conditions. In particular, specialisation in certain branches of agriculture, like potato growing in Českomoravská vysočina, for example, was made impossible by these factors. In like manner, the region under study here was characterized by utterly different economic and possession structures stemming from its historical development. Unlike the borderland of Českomoravská vysočina, where land ownership had never been so concentrated, the extent of the Schwarzenbergs' possessions and property in this region made them major economic players. The origins of the manorial estate lay in the tradition of the feudal demesne, which had always been run as an economically self- sufficient unit. Accordingly, its individual parts had to be made to work together. For the same reason, there was no specialisation in individual sectors or areas of production. Even after the abolition of serfdom, the owners of big estates worked at improving the balanced development of each field.

In the early nineteenth century, Czech agriculture was still based on traditional, mostly extensive methods of farming. Suitable conditions for further development were created by the abolition of serfdom in 1848-1849. Abolition on the one hand motivated small farmers by granting them agricultural land, and on the other, it deprived the largest landowners of a cheap work force and its equipment. In the region under study, intensification of production was the key to the profitability of agricultural enterprises. The Schwarzenberg manorial estates, such as the one in Třeboň, had plenty of capital; above all, they were professionally managed,

[17] SOA Třeboň, Hanus's inspection fund, inv. no. 38.

and disposed of enough land for machinery and labour to be utilized to the full. The manorial estates could thereby afford to be the first to introduce new sowing procedures, which called for intensive soil cultivation and regular fertilization. High soil fertility was also supported by technically demanding improvements and drainage systems. The Schwarzenberg estates made use of improved plants and breeds, put the latest technical innovations into practice without delay and bought machinery at the cutting-edge of European and worldwide technological progress. Thanks to the exceptional Jan Adolf von Schwarzenberg and his extraordinary employees, František Horský and his successor Emanuel Hanuss, Bohemia quickly became aware of the more developed agriculture of Western European countries, and adopted their rational experimentation, thereby enriching local farming methods. These factors were the base for the successful intensification of agricultural production.

The second half of the nineteenth century witnessed dynamic development in Bohemian agriculture. The volume of vegetal production, expressed in grain units, increased by more than 100 per cent during the period 1848-1914; beef production rose almost threefold, while pork output went up by even more (Jakubec and Jindra, 2006: 114-115). The increase can be credited to the eventful year of 1848, since its outcomes enabled capitalist production relations to make their way into the agricultural sphere. High compensation for freeing subjects' farms from feudal obligations not only contributed to financing modern agricultural mechanization, but also to the development of the food-processing industry, which in the 1870s helped overcome the onerous consequences of the economic crises that followed on agricultural overproduction (Lom, 1979: 145-162).

Bibliography

Albert, Eduard (1970), 'Ekonomika moravského zemědělství v druhé polovině 19. století', *Prameny a studie*, 8, 497 p.

Bílek, Tomáš V. (1893), *Statky a jmění kollejí jesuitských, klášterů, bratrstev a jiných ústavu v království Českém od Josefa II. zrušených*, Prague, Frant. Bačkovský, 708 p.

Hanuss, Emanuel (1860), 'Die Viehzucht ist kein nothwendiges Nebel, vielmehr das einziges Mittel zur Hebung des landwirtschaftlichen Wohlstandes', *Centralblatt für die gesamte Landeskultur*, 14, p. 105-109 and 16, p. 121-126.

Hanuss, Emanuel (1875), 'Beschreibung der landwitschaftlichen Regen –Wirtschaft auf der fürstlichen Schwarzenbergischen Domaine Wittingau', *Prager Festschrift,* 8, p. 27-59 and 79-91.

Hanuss, Emanuel (1877), 'Die Dampfboden-Cultur', *Prager landwirtschaftlichen Wochenblatte,* 6 (December), p. 23-31.

HESKE, František (1909), *Popis knížecího schwarzenberského svěřenského panství třeboňského,* Prague, 142 p.

HORSKY, František (1862), *Horského polní kázání,* Prague, 64 p.

HORSKY, Franz (1847), *Die Fruchtwechselwirtschaft,* Prague, 93 p.

HORSKY, Franz (1860), *Rolnická polní kázání,* Prague, 72 p.

HORSKY, Franz (1861), *Die Verbreitung der Fruchtwechsel-Wirtschaft,* Prague, 96 p.

HULE, Miroslav (2000), *Rybníkářství na Třeboňsku,* Třeboň, Carpio, 250 p.

JAKUBEC, Ivan and JINDRA, Zdeněk (2006), *Dějiny hospodářství českých zemí od počátku industrializace do konce habsburské monarchie*, Prague, Karolinum, 471 p.

KADLEC, Jaroslav (2004), *Klášter augustiniánských kanovníků v Třeboni,* Prague, Karolinum, 340 p.

KALNÝ, Adolf (2001), 'Priorita třeboňského panství v drenážování', *Archivum Trebonense,* 8, p. 131-133.

KALOUSEK, Josef (ed.) (1910), *Archiv český xxv,* Prague, 800 p.

KAZBUNDA, Karel (1929), *České hnutí roku 1848,* Prague, Klub Historický, 434 p.

KOČÍ, Josef (1957), 'Příspěvek k rolnické otázce v Čechách v r. 1848', *Československý časopis historický,* 5, p. 73.

KRAFT, Quido (1872), *Ein Grossgrundbesitz der Gegenwart. Monografische Skizze der Besitzungen des Fürstenhauses Schwarzenberg in Böhmen als Beitrag zur Frage der Selbsrverwaltung oder Verpachtung von grossgütern in Oesterreich,* Wien, 344 p.

KŘIVKA, Josef (1989), *Výnosy hlavních zemědělských plodin v letech 1870-1913,* Prague, Prameny a studie Zemědělského muzea v Praze. Sv. 32, 356 p.

KROFTA, Kamil (1949), *Dějiny selského stavu,* Prague, Jan Laichter, 427 p.

KUTNAR, František (2005), *Malé dějiny brambor,* Prague, Etnologický ústav AV ČR, 216 p.

LOM, František (1979), 'Důsledek kapitalistických krizí v zemědělské malovýrobě', *Prameny a studie,* 20, Prague.

LOUDIL, Lumír (1980), 'Vývoj živočišné výroby na území ČSR v období kapitalismu druhé poloviny 19 století', *Prameny a studie,* 22, p. 1-142.

MACZAK, Anton (ed.) (1981), *Encyklopedia historii gospodarczej Polski do 1945 roku,* Warzsawa, Wiedza powszechna, 629 p.

MATLAS, Pavel (2006), 'Proměny zemědělské výroby na Třeboňsku v 19. století', in Václav DUDÁK (ed.), *Třeboňsko – příroda, historie*, Prague, p. 246-262.

MENCEL, Tadeusz (1987), 'Chłopska własność I posiadanie ziemi przed uvłaszczeniem w Królestwie Polskim', *Roczniki Dziejów Spolecznych I Gospodarczych.*

PÁTEK, Jaroslav (1971), 'Racionalizace zemědělské výroby mechanizací v českých zemích', *Prameny a studie,* 10, 244 p.

ROUBÍK, František (1959), 'K vyvazení gruntů v Čechách v letech 1848-1853', *Sborník archivních prací,* 9, p. 160-220.

ŠIMUNEK, Robert (2005), *Správní systém rožmberského dominia v pozdně středověkých Čechách,* Unpublished Ph. D. University of Prague, 737 p.

SOMMER, Johann Gottfried (1841), *Das Königreich Böhmen. Statistisch-topografisch Darstellung,* IX, Prague, 397 p.

ŠUSTA, Josef (1995), *Pět století rybničního hospodaření v Třeboni,* Třeboň, Nakl. Carpio, 212 p.

TEMPÍR, Zdeněk (1986), 'Historie sklizně obilovin', *Prameny a studie,* 29, p. 5-82.

URBAN, Otto (2003), *Kapitalismus a česká společnost,* Prague, Nakladatelství Lidové noviny, 323 p.

VILIKOVSKÝ, Václav (1936), *Dějiny zemědělského průmyslu v Československu od nejstarších dob až do vypuknutí světové krise hospodářské,* Prague, Publikace ministerstva zemědělství, 948 p. + 278 p.

8. Post-Communist transition in Bulgaria. Implications for the development of agricultural specialisation and farming structures

Hrabrin BACHEV

I. Introduction

Since the collapse of the Communist system in 1989, Bulgarian agriculture has been going through an unprecedented transformation from a centrally planned to a market based liberal economy. This fundamental transition had a significant effect on agricultural specialisation and farming structures (Bachev & Tsuji, 2001; Bachev, 2011; Ghosh, 2008; OECD, 2000). The unique post-communist 'Bulgarian experience' provides an extraordinary opportunity to study the evolution and factors of agricultural specialisation (Bencheva, 2005). There are no other comprehensive studies on agricultural specialisation during this most recent period of development.

The goal of this paper is to fill the gap, looking at the forms and determining factors of post-communist agricultural specialisation and farming structures in Bulgaria. We have incorporated an inter-disciplinary methodology from the New Institutional and Transaction Costs Economics based on contributions of Coase (1937), Furuboth and Richter (1998), North (1990), and Williamson (1996). First, we outline the framework for analysis of economic specialisation in transitional agriculture. Next, we present features of the Bulgarian model of farming transformation, specifying the factors that led to change in farming structures and specialisation, and their impact on agricultural specialisation. Finally, we look at case studies of governance and specialisation in dominant business-run farms, cooperatives, and unregistered farms. The study is based on official data, such as statistics and censuses, as well as original information collected from managers of what can be considered typical farms in all major regions. The survey was carried out in 2003 during the last Agricultural Census and covers 194 farms of different types (0.5 per cent of the commercial farms in the country). Thirty eight per cent of surveyed farms were unregistered 'individual, family, or group farms', almost twenty nine per cent were 'cooperatives', and one-third had the status of 'firms'.

Agricultural specialisation and rural patterns of development, ed. by Annie ANTOINE, Turnhout, 2016 (*Rural History in Europe*, 12), p. 151-173.

BREPOLS PUBLISHERS DOI 10.1484/M.RURHE-EB.5.112266

II. New Institutional Economics framework

The approach provided by the New Institutional Economics enables us to better understand the logic and driving factors of agricultural specialisation and the development of farming structures (Bachev, 2004; Bachev, 2009; Masten, 1991; Sporleder, 1992).

The potential for increasing productivity provided incentives for the division and specialisation of labour[1]. In agricultural production, specialisation could occur with particular products such as crops, livestock, fruits, cereals, milk, beef-cattle, organic honey, or with specific functions such as management, mechanization, plant protection, harvesting, security, risk-taking, marketing. Labour specialisation inevitably requires the coordination of specialized activity and exchange of products, resources and rights between individual agents. Specialisation and individual exchanges could be coordinated and stimulated by a free market, through price-movements and private deals, by decisions made within a private organization by a manager or collectively, or by a third party, such as a state authority.

If transaction costs were zero[2], then all modes of governance – through markets, contracts, family farms, partnerships, cooperatives or a nationwide hierarchical organization -- would have equal efficiency. Then the type and extent of specialisation would depend only on technological factors – the ability to economize on production costs, increase economies of scale or scope, increase productivity, and profit from mutually beneficial exchange. However, when transaction costs are positive the type of governance has a significant impact on the evolution of agricultural specialisation and exchange. For instance, when a farmer integrates mechanized services by buying a tractor and hiring a tractor specialist instead of obtaining them from the market, the economic benefits are more than just technological as he will also save on production cost. Managing activity internally, by specialisation, often has substantial transaction cost advantages, by reducing the cost of finding the best prices and suppliers, and that of negotiating conditions of exchange; it reduces market uncertainty and risk from outside dependency, and so forth.

There are a great range in the specialisation of agricultural activities within internal organizations and/or across markets. At one extreme, a high level of external transaction costs could restrict the development of specialisation in small-scale subsistence farming or bring about multi-product integration into large member-

1 Higher productivity could be also achieved through cooperation of labour.

2 Transaction costs are 'costs for protection and exchange of property rights', FURUBOTH & RICHTER, (1998: 101).

oriented cooperatives. At the other extreme, a farm entrepreneur specialized in management could carry out all the production activity, purchasing all related services (tilling, watering, fertilizing, plant protection, harvesting) from specialized markets[3]. Deciding which form of management to use for a particular type of activity depends on the comparative efficiency, the advantages and disadvantages, of alternative forms.

A rational agrarian agent would tend to choose the most efficient mode for governing relations with others – the one which will achieve maximum productivity benefits while minimizing total production and transaction costs[4]. Efficient forms for managing farming activity will vary according to an individual's own characteristics (his skills and ability) and the specific institutional, market and natural environment. For instance, in transitional economies, property rights were not well defined and enforced, individuals had little managerial experience and faced significant institutional, market, and behavioral uncertainty. In these conditions, subsistence that is, semi-market household holdings and large integrated cooperatives and agri-firms happened to be the most effective forms of farming organization (Bachev, 2004: 140). For the subsistence households, high transaction costs hampered productivity growth that might have come through agricultural specialisation and kept farm size far below what was warranted by the technological opportunities. In the case of the large units, the huge potential for transactions benefits increased internal specialisation enormously and extended the horizontal and vertical holdings of farms beyond what would be expected from technological determinants alone.

Using this new framework, we shall study the farm as a governance structure possessing consumption, production, and transaction optimisation functions (Bachev, 2004: 139). In order to explain the way in which forms of specialisation changed and how they were assigned to different farming organizations, we have to analyze the structure of agrarian transactions and associated transaction costs. Since many of the transaction costs are hard to quantify, the analysis will concentrate on their critical factors: institutional factors like formal and informal property rights and the way they are enforced; behavioural factors such as agents' bounded rationality, the tendency to opportunism, risk aversion and concern for reputation; dimensional factors like frequency, uncertainty, asset specificity, and the tendency to appropriate transactions; and technological factors.

Next, in order to assess the degree to which different farms were specialized, we have to examine how they manage the supply of labour, services, inputs, land and

---

3 This is not a hypothetical case but is typical in a good part of Japanese rice farming.

4 Often there is a trade-off between increasing productivity by more specialisation and adding governing costs to pay for coordination and exchange.

finance, and how they market their output. For instance, intensified external trade, as seen in the supply of inputs and services and the marketing of farm produce, is an indicator of increased market specialisation. Within the farm, adding more labour in order to produce its own inputs and services internally demonstrates a greater degree of internal specialisation. More important, the full efficiency of individual forms of specialisation can only be understood in the framework of governance as applied to the entire farm. For example, a livestock farm could be effectively extended by developing a new specialisation. If, for example it were to produce forage for its animals by leasing in farmland and hiring labour to raise the crops, instead of merely enlarging its stockraising speciality by buying more forage on the market, multiplying the number of animal, hiring more livestock hands or selling more livestock products.

Finally, to evaluate the true efficiency of various kinds of agricultural organization the analysis will take into account their comparative efficiency and complementarity, and also the larger household and rural economy. For instance, low productive multi-product cooperatives proved an effective form for managing specialisation and exchanges in transitional countries marked by widespread small-scale and subsistence farming (Bachev, 2004: 141). Conversely, a less effective form of agricultural specialisation such as part-time farming turned out to be essential to the economy of household specialisation in transitional conditions of high unemployment, great insecurity, and significant food costs.

III. The Bulgarian model of farming change

During the Communist period (1944-1989) Bulgarian farming was carried out on large state farms averaging thousands of hectares with up to 7,000 employees. Private ownership of major agricultural resources was abandoned along with market exchanges. State farms were responsible for supplying local communities with food, producing a broad range of farm and processed products. Most activities were governed by a Central Plan through an overcrowded multilevel hierarchical organization; up to thirty per cent of agricultural employment consisted of bureaucrats. There were many reorganizations (Table 8.1) and much experimentation with economic mechanisms devised for greater efficiency. Nevertheless, up to the collapse of the socialist model, incentives and productivity were generally lacking, while most farm resources and production structure remained under state control. Small private farms were also allowed, mostly to supply food to the farm households themselves. Despite their small size, these personal plots provided a large share of certain products – maize, potatoes, eggs, meat, honey, tobacco, fruits and vegetables (National Statistical Institute).

After 1991, farmland that had been forcibly nationalized or incorporated into cooperatives was returned to its previous owners. A complex land reform was implemented, which took almost ten years to complete. It affected 85 per cent of agricultural land and turned three-quarters of households into owners of farmland. Furthermore, the former cooperatives and other organizations built on their bases were liquidated and their assets parcelled out into individual shares. Liquidation took more than four years and made more than two million Bulgarians into owners of small stakes in the assets of former state farms.

Table 8.1. Development of farming structures in Bulgaria

Type of farms	*Communist period*					*Transition*		
	1945	*1958*	*1969*	*1985*	*1989*	*1995*	*2000*	*2005*
Private farms (000)	1,300	330	-	1,600	1,600	1,777	755.3	515.3
Average size (ha)	3.5	2.6	-	0.38	0.38	1.35	0.9	1.8
Cooperatives	35	3,202	795	-	-	2,623	3,125	1,525
Average size (ha)	n.a.	1,100	4,140			800	709.9	584
State farms	-	54	159	-	-	1,002	232	-
Average size (ha)		3,630	4,040			338	358	
Agro-industrial complexes	-	-	-	298	-	-	-	-
Average size (ha)				12,600				
Collective farms	-	-	-	-	2,101	-	-	-
Average size (ha)					1,820			
Agro-firms	-	-	-	-	-	n.a.	2,275	3,704
Average size (ha)						n.a.	296.7	249.4

Source. NSI (National Statistical Institute).

A fundamental transformation of the economy was also carried out by liberalizing markets, privatising public enterprises, and introducing EU institutions and standards. Economic reforms unleashed market competition and brought in strong incentives for private entrepreneurship. More than 1.9 million private farms emerged with provisional or entirely restored private ownership rights to lands and agricultural assets. By 1995, almost all agricultural activities, including specialisation, were governed by entirely new market and private structures. The previous model of agricultural specialisation within large state farms and a nationwide hierarchy was replaced by a capitalistic system of private entrepreneurship and free market mechanisms.

IV. Factors in the transformation of farm structures and specialisation

The specific form and pace taken by the privatization of agrarian resources and significant uncertainty during the transition had important consequences for the development of farm organization in the country.

Firstly, the long delay before agricultural resources attained full ownership held back the development of feasible forms for their effective organization. Until 2000, when sale and long-term lease markets for farmland emerged, the annual lease was the main way in which farmland could be extended. Actors in the agricultural economy could not get full return on property rights or use land ownership to help organize their various activities and dealings, for example, by using their land as loan collateral. On the other hand, the unspecific or ideal character of ownership made it possible to consolidate fragmented farmland rapidly, under the management of a small number of huge market-oriented enterprises (Table 8.2). Thus the governance of a significant share of agricultural activity and specialisation was carried out within integral organizations rather than by market competition.

Table 8.2. Share of different farms in total holdings and major resources in Bulgaria

Indicators	*Physical persons*	*Cooperatives*	*Sole traders*	*Companies*	*Associations*
Number of holdings with UAA (%)	99.0	0.3	0.4	0.2	0.05
Utilized agricultural area (%)	30.3	40.3	11.7	16.1	1.6
Average size (ha)	1.4	592.6	118.8	352.5	126.2
Number of breeders without UAA (%)	96.1	0.2	1.9	1.7	0.1
Workforce (%)	95.5	1.2	0.8	1.4	0.3
Labour input (%)	91.1	4.1	1.4	2.8	0.6

Source. MAF (Ministery of Agriculture and Forestry) (2005).

Secondly, the internal organization of available household resources into individual or family farms was an effective way of overcoming great institutional and economic uncertainty and minimizing transactions costs. During much of the transition, the market and contract trade of household capital was either impossible or very expensive – a missing markets situation with high uncertainty and risk, asymmetry of information, and a real possibility of opportunism in time of hardship. Many lost their jobs as result of the restructuring of public companies. Low payoff from outside trade was combined with an increased share of household food costs. Internal organization turned out to be the most effective way for households to protect and get a return on resources, and obtain a secure, stable income. A long tradition of cultivating personal plots from the Communist period, and the insignificant cost of acquiring specialized knowledge made development costs for privately-owned farms something everybody could manage, regardless of previous occupation. Production for self-consumption has proven to be an effective mode guaranteeing cheap, stable and high-quality delivery of food, provision of employment for family members, or a useful spare-time occupation.

That is how a huge subsistence and part-time farming sector emerged, affecting a large number of households. Micro-specialisation on a full-time basis by unemployed or retired persons, or part-time diversification into farming by people engaged in other trades, took place. Even now, those employed full-time in the sector account for 11.5 per cent of all workers in the country (National Statistical Institute, 2005). In addition, almost one million Bulgarians are involved in farming on a part-time basis as a supplementary income source. Estimates in Annual Work Units (AWU) show that agriculture comprises more than 26 per cent of overall employment (MAF, 2005). Labour contributed by part-time workers constitutes 53 per cent of AWU in the sector.

Now more than three-quarters of farms contain less than one hectare of land, half a hectare on the average. More than 97 per cent of livestock holdings are non-professional farms with few heads of livestock, but they breed 96 per cent of goats, 86 per cent of sheep, 78 per cent of cattle, and 60 per cent of pigs. More than 90 per cent of farms amount to less than two European Size Units[5] and are generally considered subsistence and semi-subsistence farms. In livestock grazing and various types of mixed farming they make up a considerable share of the Standard Gross Margin of their respective groups (MAF, 2005).

Within small farms there is no labour specialisation, or it is carried out between family members. Most farms are not specialized for the market, but rather aim at serving the diversified food needs of the household, producing a wide range of crops and livestock products. For vegetables, fruits, vines and livestock a significant portion of the overall national output is for the unit's own consumption (MAF, 2005). A large proportion of farms sell only their surpluses of major commodity products (Figure 8.1). For Physical Persons the number is higher, as less than 39 per cent of

[5] One European Size Unit (ESU) is a standard gross margin (SGM) of 1200 euro, used to express the economic size of an agricultural holding or farm. The SGM is estimated on the basis of the area used for a particular activity, or the number of head of livestock, with a regional coefficient. The sum of all SGM is then expressed in ESU by dividing the total expressed in euro by 1200 (Eurostat).

farms report that they sell products and on more than 50 per cent of them these are surpluses not consumed by households (MAF, 2005).

Thirdly, most actors in the agricultural economy found that their skills, their previous specialisation and traditions of working together, and their share of acquired agrarian resources were highly interdependent. A great part of individuals' shares in former state farms were in indivisible assets – large machinery, buildings, irrigation facilities and the like. For such, the only alternatives were liquidation by sale or consumption, or keeping them in joint or cooperative ownership. In many cases, 'new' land owners received their plots as restitution, with fruit trees, vineyards and the like, and they could in practice carry out many of their activities in a cooperative. As it happened, most land and share-holders lived away from rural areas, or had another business, were aged or lacked the skills or capital to start their own farms. In the absence of strong demand for farmland and confidence in the new emerging private formats, the only possibility was to join a cooperative. That is how more than 40 per cent of new owners came to pool their free land, assets, and labour in new production cooperatives. Similarly, most privatised state farming and livestock companies were taken over by teams of managers and registered as Shareholder Companies and Associations, retaining their previous internal and market specialisations.

Figure 8.1. Share of farms marketing products in Bulgaria

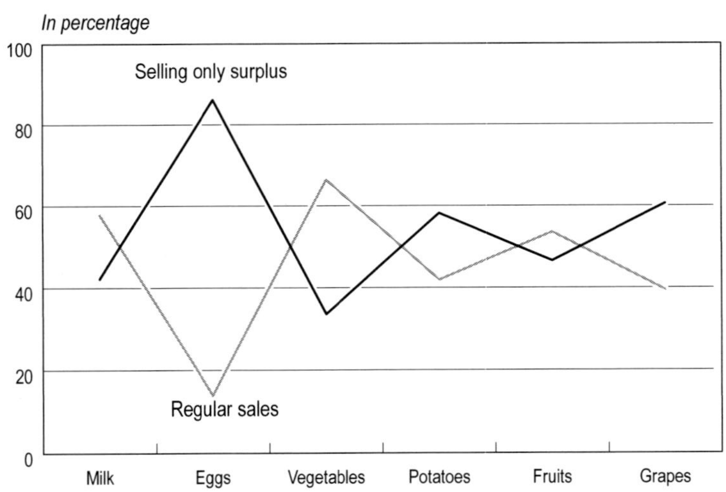

Source. Agricultural Holdings Census (2003).

Fourthly, the majority of new entrepreneurs had little or no experience in managing private or collective farms. Moreover, there has been little public support for agricultural training, advisory services and funding (Bachev, 2006: 143). That has been coupled with strong competition with heavily subsidized foreign producers in both local and international markets. As a result there have been massive failures, take-overs and cessation of commercial activity in a good number of newly-evolving farms. Between 1994 and 2005, the number of farms shrank by 72 per cent. Thus a large portion of agrarian activity and resources have gone through several governance structures or fallen out of productive use. Such unsustainability of farming structures has big implications for the extension of agricultural specialisation.

Fifth, incentives for long-term investment in specialized and specific capital in most farms have been low (Figure 8.2). That has deterred development of specialisation both within farms and across markets. For a long time, a significant portion of agricultural resources were managed by ineffective temporary structures (Privatisation Boards, Liquidation Councils and the like) with no interest in long-term productivity. High instability and uncertainty do not encourage the sustainability and investment activity of commercial enterprises (Bachev, 2006: 137). Moreover, much farm-related investment is highly specific and can hardly be funded by outside credit or equity sales (Bachev, 2006: 137). The majority of small-scale commercial farms are run by older entrepreneurs with a short-term business and investment perspective[6]. Almost all the farmland for large farm businesses has been leased on provisional contracts. While there has been strong investment in mobile material assets, no long-term money has been put into improving land productivity by renovating orchards, vineyards, irrigation facilities and restoring nitrogen, phosphate and potassium levels. Most cooperatives have been mismanaged or experienced serious funding problems because of their members' divergent investment preferences (Bachev, 2006: 142). By contrast, subsistence farming does not require significant investment, since household demand is stable.

V. Impact of farming transformation on agricultural specialisation

Post-communist transformation of agriculture has increased the importance of farming in the national economy both in terms of its contribution to the Gross Domestic Product and to employment (Figure 8.2). Compared to 1989, agricultural employment increased by twenty per cent as its share in overall employment doubled. Thus overall agricultural specialisation expanded as more people found work and income in this sector of the national economy. On the other hand, 75 per cent of those employed in farming are engaged on a part-time basis, showing the predominance of

[6] The proportions of managers over the age of 45 and 65 are 85 and 40 per cent respectively (MAF, 2005) .

primitive, rather than complete, forms of agricultural specialisation with less labour specialisation and fewer exchanges.

Figure 8.2. Share of Bulgarian agriculture in national economy

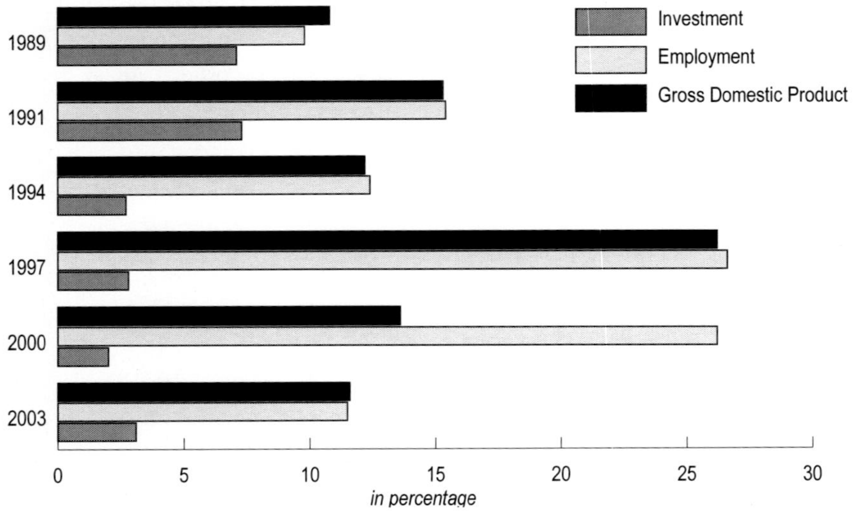

Source. National Statistical Institute.

Figure 8.3. Dynamics of major agricultural productions

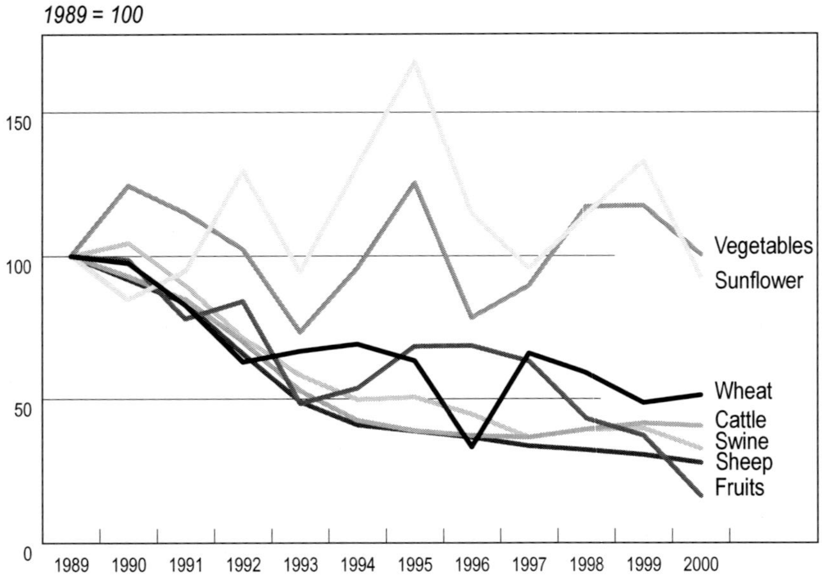

Source. National Statistical Institute.

Figure 8.4. Composition of gross agricultural product in Bulgaria

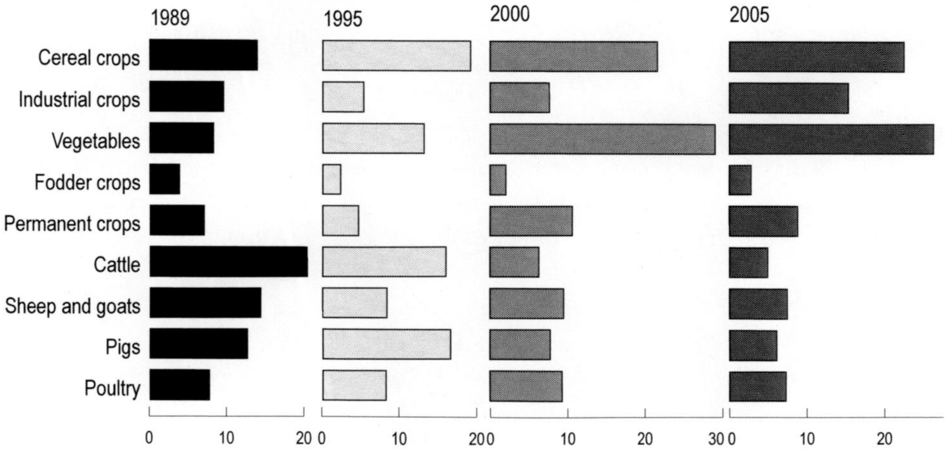

Source. National Statistical Institute.

Figure 8.5. Share of farms with different specialisation and contribution to standard gross margin of agriculture

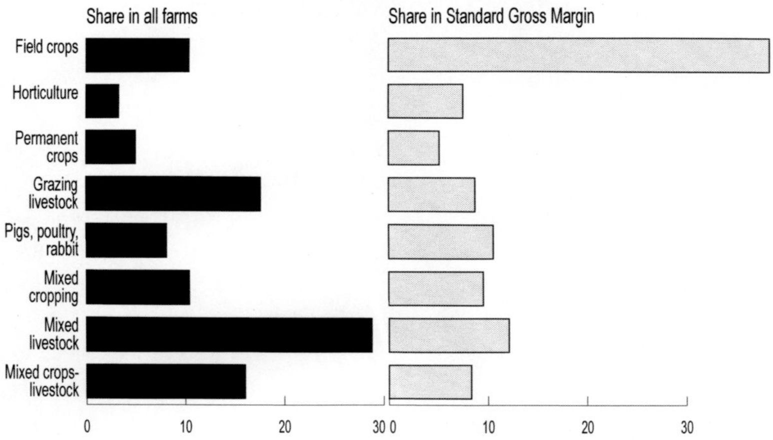

Source. Agricultural Holdings Census (2003).

Post-communist market adjustment and farm adaptation has been associated with an alteration of production structure and a general decline in agricultural output (Figure 8.3). There has been a significant change in general product specialisation and importance of different sub-sectors (Figure 8.4). Livestock lost its dominant pre-reform share, falling to 26 per cent of the Gross Agricultural Product (GAP). While poultry maintained its position, the proportion of cattle, sheep and goats, and pigs dropped considerably. At the same time, the share of vegetables grew significantly: formerly in sixth place, it now rose to become the top contributor to GAP. Cereals and industrial crops also rose in scale to become the second and the third most important productions.

The newly evolving farming structure is characterized by entirely different types of specialisation from the pre-reform period. A sizable proportion of farms produce many different products. Only in cereals, oil plants, aromatic and medicinal crops is production more specialized. The fraction of large-scale specialized livestock farms is insignificant. While various types of mixed farms and primitive livestock-grazing farms comprise the bulk of the farms, the greater part of the Standard Gross Margin (SGM) of the sector (Figure 8.5) is produced on different kinds of specialized farms. Larger commercial farms (bigger than two ESU), with different types of specialisation, contribute most to the SGM in all specialisation groups. A tiny number of all farms in permanent crops (0.3 per cent), field crops (1.6 per cent), pigs and poultry (0.2 per cent), horticulture (0.4 per cent), and mixed cropping (0.25 per cent) are larger than 100 ESU, but they produce a significant share of the SGM in these groups – 63 per cent, 54 per cent, 54 per cent, 26 per cent, and 28 per cent respectively (MaF, 2005).

VI. Management and specialisation of farm businesses

Our survey has found that agro-firms are commonly large specialized enterprises. Most of these firms were set up during the first years of transition by entrepreneurs of the younger generaion, as family or partnership businesses. Specific management skills and social status, and the combination of partnership assets such as technological knowledge, business ties and available resources made it possible for farms to grow quickly in size by bringing together on an enormous scale, resource management or ownership, specialisation and exploration of economies of scale and scope within modernized entreprises (Bachev, 2006: 136). Institutional uncertainty, vague rights to assets, personal relations and virtually or entirely integrated modes were widely used to overcome transaction difficulties. The number of agro-firms has increased twenty-fold since 1990 and doubled since 2000, as their share in overall resources increased (National Statistical Institute).

Business farms are profit-orientated organizations, and farmers have great incentives to invest in farm-specific capital – human, material and intangible – because they are sole owners of the residual rights and benefits of the farm. Owners are family members or close partners, so the internal transaction costs of coordination, decision- making, and motivation are low. The firm is the preferred organizational style because it helps overcome the problems of association by enabling joint ventures with outside capital or the resolution of conflicts through the courts. Farming firms can branch out into farm-related or independent businesses. They can develop firm-specific intangible capital like advertising, brand names and public confidence; they can form subsidiary companies; they can trade and transfer across generations; and they are able to overcome existing institutional restrictions to direct foreign investments in farmland and to trade in cereal, vine and dairy products. The status of a firm confers explicit rights to take part in particular types of transaction such as licensing, privatisations and public programs.

The large scale and good reputation of farm businesses make them preferred partners in input supply and marketing deals. Transactions with the same partners recur frequently, limiting information asymmetry and opportunistic behaviour, and fostering mutual trust and other mechanisms that facilitate and reduce the cost of relationships, such as planning, forms of adjustment and payment, guaranty schemes and dispute-resolution mechanisms. Agro-firms have huge negotiating power and economic, political and other mechanisms to enforce contracts effectively. They are very well placed to collect market information, search out optimal partners, employ expertise and innovation, meet special collateral requirements and bear the risk and cost of failures. They can increase economies of scale and scope in production and management: for instance, by 'packaging' loans for multiple projects and interlinking input with know-how, supply, credit and marketing. They are also in a position to invest considerable relation-specific capital like information, expertise, reputation and lobbying into dealing with funding institutions or the agrarian bureaucracy or market agents on the national or international level.

When courts and contracts fail to guarantee deals, all critical farm transactions are controlled and protected internally. Farm-specific assets such as essential machinery, vineyards, orchards, animals, processing facilities, and adjoining land are all safeguarded by ownership. Low-cost, standard contracts are widely used to lease land from tens to hundreds of landowners.

'Critical' transactions are integrated through extensive labour employment (Figure 8.6). The choice of in-house provision by internal specialisation and organization of activity, over buying specialized inputs and services in the market is determined by the economic rationale of transaction costs – the high farm specificity and/or uncertainty associated with certain transactions. Moreover, core labour provided by specialists or mechanics is hired on a permanent basis; special forms of compensation such as output-based remuneration, interlinking of housing and services, social disbursements and paid holidays are also used to enhance motivation. Furthermore, in large business farms it is typical to have a strong internal division and specialisation of labour both according to function (management, production, marketing and security) and by area (crops, livestock and services). Nevertheless, in farming, hired, non-family, labour incurs high transaction costs for directing, supervising, preventing opportunistic behaviour and dealing with disputes. So large business farms are mostly specialized in less labour-consuming, highly-intensive and standardized productions – cereals, sunflowers, aromatic and medicinal plants, poultry and pigs for instance – where control of labour is easy or when output-based compensation may be effectively applied (Tables 8.3 and 8.4). These farms are also major providers of mechanized services to other farms.

Figure 8.6. Governing of labour supply in Bulgarian farms

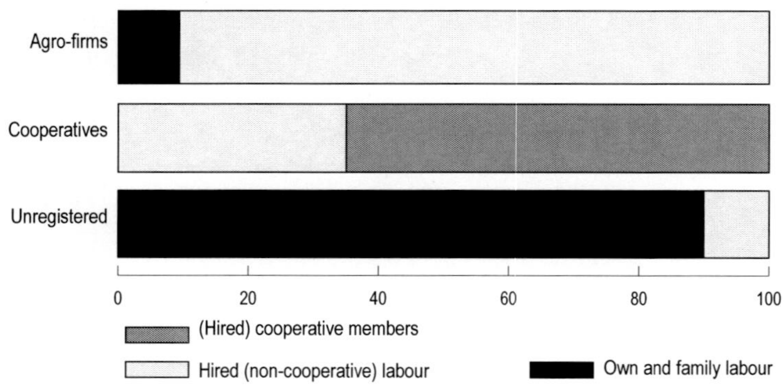

Source. Personal interviews with farm managers.

Table 8.3. Crop productions by farm type

Productions	*Unregistered*		*Cooperatives*		*Sole traders*		*Companies*		*Associations*	
in % for each type of crops	*Farms*	*Area*	*Farms*	*Area*	*Farms*	*Area*	*Farms*	*Area*	*Farms*	*Area*
Cereals	51.1	59.5	93.0	59.7	54.3	62.3	55.1	61.1	51.1	56.0
Dry pulses	13.0	0.8	5.7	0.6	5.9	0.6	3.6	0.5	5.9	1.1
Oil crops	5.1	17.8	74.4	28.6	28.9	29.7	41.6	27.0	28.2	26.0
Textile crops	0.0	0.0	1.0	0.2	0.1	0.0	0.4	0.2	0.5	0.2
Aromatic & medicinal plants	0.5	1.7	17.6	2.5	4.8	1.9	12.8	4.4	6.7	5.6
Forage	28.6	7.5	29.6	3.1	21.2	1.2	14.0	1.7	23.4	7.5
Vegetables	49.2	5.5	8.3	0.2	29.2	0.5	11.9	0.4	21.2	0.4
Fruits	8.7	2.0	14.7	0.9	11.4	0.3	12.2	0.7	35.2	9.5

Source. MAF (2005).

Table 8.4. Livestock productions by farm type

Productions	*Total*	*Unregistered*	*Cooperatives*	*Sole traders*	*Companies*	*Associations*
Cattle	31.8	32.0	13.1	18.8	11.3	22.8
Buffalo	0.4	0.4	0.2	0.5	0.7	0.8
Goats	40.4	40.8	0.5	14.7	2.9	6.5
Sheep	35.7	36.0	5.0	17.1	6.3	13.7
Pigs	41.9	42.1	3.6	29.9	15.1	20.4
Poultry	74.3	74.8	1.2	43.2	13.2	15.3

Note. % of farms with each type of livestock

Source. MAF (2005).

In-house rather than outside procurement – 'making' rather than 'buying' – is common for essential services and inputs in the main (Tables 8.5 and 8.6). It prevents the domination of a single supplier, or a 'missing' market situation. In the case of high asset interdependency with downstream partners – as when there is product specificity or quality and/or quantity dependency, a reciprocal supply of 'inputs against marketing' is applied.

Table 8.5. Governing of service supply in Bulgarian farms

Service type	*Modes*	*Unregistered*	*Cooperatives*	*Agro-firms*
		percent of farms		
Technological knowledge and advises	Own supply	24	49	65
	Own cooperative	5	7	15
	Market supplier	13	10	25
Mechanization services	Own supply	18	85	60
	Own cooperative	22	0	18
	Market supplier	15	15	28
Spreading chemicals and pesticides	Own supply	40	65	60
	Own cooperative	15	7	12
	Market supplier	12	25	28
Veterinary services	Own supply	20	60	40
	Own cooperative	5	0	0
	Market supplier	40	40	60

Source. Personal interviews with farm managers.

Table 8.6. Governing of inputs supply in Bulgarian farms

Inputs type	*Supplier*	*Unregistered*	*Cooperatives*	*Agro-firms*
			percent of farms	
Chemicals	Own production	17	0	0
	Own cooperative	10	5	15
	Market supplier	55	95	90
	Buyer of farm output	24	13	33
Seeds and seedlings (crop farms)	Own production	47	53	33
	Own cooperative	3	15	23
	Market supplier	50	32	45
	Buyer of farm output	4		44
Forage (livestock farms)	Own production	55	65	50
	Own cooperative	0	0	35
	Market supplier	45	35	15
	Buyer of farm output	9	6	53
Machinery	Own production	12	13	0
	Own cooperative	20	17	46
	Market supplier	68	70	54
	Buyer of farm output	15	0	19
Livestock	Own production	37	50	28
	Own cooperative	21	31	33
	Market supplier	42	19	39
	Buyer of farm output	40	17	13

Source. Personal interviews with farm managers.

Our survey has found out that funding is secured through an effective combination of equity and debt and of public and hybrid modes. Standard activities or assets are financed by bank loans, which are easily arranged. Alternatively, farm-specific investments can be financed through private modes – the firm's own resources, personal loans or co-investment. Special forms of contract can be arranged to ease funding difficulties, caused for example by delayed payments for inputs or a shortage of working capital, by employing zero-interest loans or loans in kind, or to facilitate mutually-dependent relations with buyers or suppliers, interlinking credit with inputs and marketing or leasing, or accepting outside investment in long-term assets, by 'hostage taking' or joint ownership. Agro-firms are also quite successful in getting access to recently-introduced public support programs, developing good proposals, meeting formal requirements, dealing with complicated paper work, and organizing the selection of projects, as with the 2001-2006 EU Special Assistance Programme for Agricultural and Rural Development or the 2007-2013 National Plan for Agrarian and Rural Development.

Table 8.7. Governing of marketing in Bulgarian farms

Output	*Modes*	*Unregistered*	*Cooperatives*	*Agro-firms*
			percent of farms	
Grain	Own cooperative	9	7	9
	Another farm/firm	50	85	75
	Processor	25	39	37
	Retail	6	7	16
Vegetables	Own processing	0	0	15
	Another farm/firm	24	24	35
	Wholesale market	6	5	15
	Processor	38	66	30
	Retail	12	0	6
Fruits and grape	Own processing	15	7	19
	Own cooperative	24	7	9
	Another farm/firm	48	39	32
	Wholesale market	0	22	22
	Processor	15	36	25
	Retail	6	0	0
Meat	Own processing	0	10	15
	Another farm/firm	65	71	80
	Processor	29	43	30
	Retail	15	36	20
Milk	Own processing	0	10	15
	Another farm/firm	42	43	40
	Processor	51	64	45
	Retail	19	0	15

Source. Personal interviews with farm managers.

In the marketing of farm output and services, whether in a wholesale market or with a business employing market agents, a classical cross-market trade prevails (Table 8.7). The main part of farm's product has a standardized, or commodity, character. Market prices and competition effectively governs relations with partners. When output is highly specific to a particular buyer, whether processor or retailer, because of technology, quality, time of delivery or site-specificity, then delivery contracts with the respective partner are employed to tailor or protect transactions. Here division of labour and specialisation in a vertical chain is governed by private order rather than classical market competition. Intra-firm processing and retailing is also practiced by some farms. That further extends farm boundaries through internal organization and specialisation of interdependent activities. Larger operational size and frequency of transactions provide an economic opportunity for internal exploration of interdependent assets in farming, processing and retailing. Vertical integration helps protect dependent investments and payoffs from marketing processed products – obtaining full profit, brand-name trade and lessened market dependency through advantages such as easy storage and transportation and so forth.

VII. Governance and specialisation of production cooperatives

Co-operatives are the biggest farms in terms of land and labour management (Table 8.2). Co-ops mainly concentrate on cereals, oil and forage crops, orchards and vineyards. They are also key service providers for their members and for the rural population.

The co-operative was the single most effective form of organization as long as there were no settled rights on main agrarian resources and/or in the absence of inherited high interdependence of available assets as were found on restituted farmland, individual shares acquired in the actives of former co-operatives, or narrow labour specialisation (Bachev, 2006: 141). Moreover, most co-operatives developed side-by-side with small-scale and subsistence farming. Their 'non-profit' character and strong membership, rather than market, orientation attracted many households. As for production, the co-op was seen as an effective, cheap and stable way of supplying highly specific inputs and services to individual farms, such as fodder, and machinery, and the storage, processing, and marketing of output, as well as a source of food for households. The co-operative has been mostly preferred to other formal types of collective, such as the firm. Co-ops were started up by entrepreneurs of the older generation and tradition played a role in this. Moreover, this mode allows individuals an easy, inexpensive entry and exit, retaining control over a major resource – land-- and allowed memebers to share democratically in management and to keep control over it. In addition, co-operative forms provide some important tax advantages such as exemption from sale transactions with members, and receipt of rent in kind; it is also possibile to organize transactions that are not legitimate for other organizational modes, like nation-wide credit supply, marketing and lobbying.

Larger operational size allows co-operatives a great opportunity to use resources efficiently: labour, through the division and specialisation of labour; farmland, by cultivation in big consolidated plots using effective crop rotation; and material assets, by exploring economies of scale or scope with the help of large machinery.

In addition, co-ops have a greater chance of minimizing market uncertainty by 'risk pooling', advertisement, storage¸ and processing and marketing. It is easier for them to organize critical transactions like access to credit and public support programs. They can negotiate positions in input supply and marketing, ease land consolidation through lease-in and lease-out deals; implement technological innovations, and invest in intangible capital like reputation, labels, and brand names.

Co-operative activities are not difficult to manage since internal demand from members for output and services is known and 'marketing' secure. In addition, co-ops concentrate on a few highly standardized, mass products with a stable market and profitability. All these assist financing; members commonly commission advance funding of their activities, and production of universally-consumed commodities is easily financed by public programs or commercial credit. Furthermore, co-ops offer low-cost long-term land leases. That is often coupled with simultaneous lease-out deals as a specific mode for cashing out the co-ops' output or facilitating relations between landowners and private farms. Co-ops regularly integrate critical services with the supply of inputs. They commonly pay for labour on the basis of output; this restricts opportunism and minimizes internal transaction costs. In addition, co-operatives provide employment for members who otherwise would have no job opportunities – housewives, pre-retired and retired persons. They are preferred employers since they offer higher job security, social payments, paid holidays, career prospects, and the like. Marketing risk is governed by effective delivery contracts or fitted into in-house processing. In a situation of 'missing markets' in rural areas, the co-operative mode is also the only form for organizing important activities such as bakeries, wholesale, retail trade etc. In return for considerable transactions benefits, most co-op members accept lower than market returns on their resources – lower wages, inferior or no rent for land and dividends for shares.

There have been some adjustments in the size, membership numbers, and production structure of co-ops. A number of them have moved toward a 'new generation' style of corporate governance, sharing their profit-making goals, closed-membership policies and joint-ventures with other organizations. At the same time, co-operatives have been seen to possess certain disadvantages as a form for farm organization. Wide participation makes individual and collective control over management very difficult and costly and increases the chances of mismanagement, resulting in on-the-job consumption or deals that are unprofitable for members. Besides, the investment preferences of the diverse groups of members can differ: older and younger, working and non-working members, or large and small shareholders have different perspectives because co-operative shares cannot be traded: this is known as the 'horizon problem'. Since most members are old, small shareholders, and non-permanent employees, the incentives for long-term investment in co-operatives are very low. Finally, many co-ops fall short in adapting to the divergent service needs of members and in exploring the potential for inter-co-operative modes. Accordingly, co-operatives prove less efficient in the long term in their market, contract and partnership arrangements. Almost forty per cent of them have gone bankrupt or ceased to exist in last five years.

VIII. Governing and specialisation of unregistered farms

The majority of commercial farms are unregistered farms[7], and mainly engaged in labour-intensive productions crops such as vegetables, tobacco, vineyards, berries, melons, flowers, mushrooms, medicinal and aromatic crops, livestock, sericulture and beekeeping, and the output of natural meadows. Those are predominately individual or family holdings, and farm size is determined exclusively by available household resources – farmland, labour, finance. Internal administrative costs are insignificant since transactions take placebetween family members. These people share common goals, enjoy high confidence, and generally do not cheat on each other; in the case of one-person farms, such costs are of course nonexistent. Some production activities admit a degree of collective organization, which gives some scope for exploring economies of scale or makes it possible to farm on a part-time basis. Such collective organization is cost-effective, as transactions are uncomplicated and easily controlled. Group members are usually close friends, neighbours, or relatives; mutual trust and self-restriction of opportunism govern their relations.

Commercial farmers own all the residuals in the form of income, so they have strong incentives to adapt to market demand and increase productivity by intensifying labour and investing in human and material assets. They cannot readily increase the size of their farms by bringing in more labour or services from outside as in labour-intensive, spatially dispersed production, the costs of directing, monitoring, and bargaining tend to be very high. The high cost of preparing project proposals, filling in formal paperwork, meeting requirements for ownership, co-financing and the like, and 'arranging' funding all put external financing of farming via debt, equity sale, or preferential public programs beyond reach. Thus low internal investment capacity, whether from savings or profits, limits the potential for effective farm enlargement and productivity growth through intra-farm division and labour specialisation, mechanization, application of chemicals and innovation. In general, primitive technologies and poor environmental and animal welfare standards prevail (Bachev, 2006: 143; Bachev, 2009: 29). As many as 40 per cent of surveyed farms report not using essential services at all. Commercial farms lease low-cost, outside land to

[7] Persons' unregistered farms meeting certain criteria such as a minimum farmland area, number of animals, etc.).

increase economies of scale on their existing assets. Outside supply of indispensable inputs and services such as seeds, chemicals and veterinary work do not increase costs significantly costs since they are occasional and standardized because of their low specificity and large range of suppliers. By contrast, joint ownership formats such as co-operative and group farming can easily secure highly specific feed supplies for animals and mechanized services.

The individually-owned farm enterprise has proven a sure means of providing full or part-time employment for family members. Family organization is also an effective form for intergenerational transfer of farm-specific intangible assets like know-how and reputation. However, small-scale commercial farms have little ability to meet institutional and market restrictions, to face up to risks, and to protect themselves against natural and market hazards. Many of them face great transaction difficulties in marketing their output. Most often they are not preferred partners for big buyers because of the small volume and less standardized character of their output, and the impossibility of verifying the quality of their products because of the high costs of laboratory tests, certificates and so on. But official wholesale markets have also been inaccessible to these farms because of their great distance, high fees, and their volume, special preparation and certification requirements. Besides, farms are frequently victims of unfulfilled contract obligations, when their payments are not received or delayed, and of severe market price fluctuation, monopoly situations, missing markets and so forth. In order to protect transactions and avoid unwanted exchanges they commonly employ primitive forms of risk minimization such as investment in more universal but less productive, less profitable assets, product diversification, informal cash and carry deals and direct retail marketing. Nevertheless, some smallholders have managed successfully to join modern market chains with vertical partners like processors, retail chains and exporters[8].

Except among tobacco producers, it has proved hard to develop effective collective organization of risk sharing, price negotiation, marketing and lobbying for public support. High transaction costs, 'free riders', opposing interests of individual farmers (old versus young operators for instance), larger or smaller, specialized or diversified farms, and the low reputation and inefficiency of new emerging farmers associations have all been responsible for this. Most small commercial farms are vulnerable and ill-protected from outside institutional, market, and natural disturbances. Most of them have little ability to face severe market competition, and to meet fast-evolving institutional restrictions such as those imposed by the EU, or to bear the risk and safeguard themselves against natural and market hazards by purchasing insurance, diversifying or cooperating. All these factors make individual farms vary consderably

---

8 A case of an effective integration of small dairy farms is presented by Bachev, Manolov, 2007.

between different sectors, regions, and over time. What is more, farms in entire sub-sectors such as dairying have been unable to adapt to the new market and institutional order associated with EU integration. Consequently, land tends to be managed more and more by bigger farms, and small farms tend to decrease in number, often restricting or ceasing their commercial activities altogether. (Bachev, 2009: 33).

Last but not least, unlike other forms of organization, the life cycle of the single-person family farm is greatly determined by the age of the entrepreneur. Older farmers are little inclined to invest on a long-term basis in specialized assets for increasing sustainability, because there is no secondary market for farm-specific assets like investments in human capital, good reputation, know-how and organizational modernization. So a good number of small commercial farms, most with older farm managers and labour, operate at a low level of sustainability.

IX. Conclusions

The post-communist transformation of Bulgarian agriculture has allowed us to determine the factors, modes, and extent of farming specialisation in a fast changing institutional and market environment. We have demonstrated that the particular type of institutional modernization, pace of market adjustment, and structure of transaction costs are responsible for the evolution of new and specific forms of farming organization and specialisation, quite different from other European countries. The Bulgarian transformation created a large subsistence and part-time farming sector, with large-scale co-operation in production, concentrating many resources in a few business enterprises, widely employing vertically integrated modes and private order, relying on special contracts, informal modes and personal relations. Along with this went numerous missing markets and failures, backward specialisation and technological development.

During the transition the overall 'agricultural specialisation' of the population expanded as more people became involved in farming than in the Communist period. Moreover, central planning and public organization were entirely replaced by market and private governance of agricultural specialisation and exchanges. Unlike the Communist period, much agricultural activity is now carried out in small-scale, semi-market and part-time farms with primitive forms of internal and market specialisation. These farms dominate entire sub-sectors like horticulture, tobacco, and cattle, sheep, and goat breeding. At the same time, a major part of agricultural activity in permanent crops, cereals, oil plants, poultry and pig production has been concentrated in large organizations -- cooperatives and agro-firms -- with complex internal and high market specialisation. What is more, agricultural production has declined considerably, while product specialisation and the relative importance of sub-sectors has also changed. Although livestock production has lost its dominant pre-reform share, shrinking to a quarter of GAP, vegetable, cereal, and industrial crops have expanded considerably.

Bibliography

BACHEV, Hrabrin (2004), 'Efficiency of Agrarian Organizations', *Farm Management and Rural Planning,* Fukuoka, Kyushu University, 5, p.135-150.

BACHEV, Hrabrin (2006), 'Governing of Bulgarian Farms. Modes, Efficiency, Impact of EU Accession', in Jarmila CURTISS, Alfons BALMANN, Kirsti DAUTZENBERG & Kathrin HAPPE (eds), *Agriculture in Face of Changing Markets, Institutions and Policies: Challenges and Strategies*, Halle, IAMO, p. 133-149.

BACHEV, Hrabrin, TSUJI, Masao (2001), 'Structures for Organization of Transactions in Bulgarian Agriculture', *Journal of the Faculty of Agriculture of Kyushu University,* 46, 1, p. 123-151.

BENCHEVA, Nelly (2005), 'Transition of Bulgarian Agriculture. Situation, Problems and Perspectives for Development', *Journal Central European Agriculture,* 6, 4, p. 473-480.

COASE, Ronald (1937), 'The Nature of the Firm', *Economica,* 4, p. 386-495.

FURUBOTH, Eirik, RICHTER, Rudolf (1998), *Institutions and Economic Theory: The Contribution of the New Institutional Economics*, Ann Arbor, The University of Michigan Press, XVIII-653 p.

MASTEN, Scott (1991), 'Transaction-cost Economics and the Organization of Agricultural Transactions', *NC-194 World Food Systems Project Symposium,* Chicago, p. 17-18.

MAF (2005), *Agricultural Census in Bulgaria 2003. Results,* Sofia, Ministry of Agriculture and Forestry.

NORTH, Douglass (1990), *Institutions,* Institutional Change and Economic Performance, Cambridge, Cambridge University Press, VIII-152 p.

OECD (2000), *Review of Agricultural Policies in Bulgaria,* Paris and Sofia, OECD.

SPORLEDER, Thomas (1992), 'Managerial Economics of Vertically Coordinated Agricultural Firms', *American Journal of Agricultural Economics,* 74, 5, p. 1226-1231.

WILLIAMSON, Oliver (1996), *The Mechanisms of Governance*, New York, Oxford University Press, XII-429 p.

9. The impact of the British food production campaign (1939-1945) on agricultural specialisation and rural patterns of development: a reappraisal

John MARTIN[1]

I. Introduction

The state directed food production campaign of the Second World War which was eulogised by K. A. H. Murray's in his magisterial contemporary account Agriculture as 'a success story... far beyond the calculations and estimates of pre-war planners' has, until recently, received scant attention by historians (Murray, 1954: 340). This omission is rather surprising given that it is commonly acknowledged that the efforts of British farmers not only saved the country from starvation but also instigated a productivist agricultural revolution characterised by specialisation and the commercialisation of farming methods. Despite the fact that Britain was highly dependent on imported food, the efforts of the farming community ensured that the population was neither starved into submission, nor forced to endure the food shortages and malnutrition that helped to undermine the morale of many other combatants engulfed in the world's greatest ever military conflict (Milward, 1977: 253-9).

This wartime transformation has been hailed as the precursor of the post-war revolution in British agriculture under the influence of state directives, high levels of output, economic efficiency and unprecedented increases in labour productivity were achieved as a result of specialisation in a variety of forms. The post-war success of British agriculture can be attributed primarily to the fact that British farming exploited its comparative advantages by concentrating on producing agricultural commodities in those regions to which they were best suited by dint of climate and topography. Economies of scale were employed in terms of the size of farms and

[1] The author would like to acknowledge the generous support of the the Faculty of Art, Design and Humanities, De Montfort University; Leverhulme Foundation in granting him a fellowship to undertake the research ; the British Agricultural History Society for funding the publication of *The Frontline of Freedom: British Agriculture in the Second World War* (2007), and to the co-editors Brian Short and Charles Watkins, and the Research Fellowship provided by the Museum of English Rural Life, Reading University.

Agricultural specialisation and rural patterns of development, ed. by Annie ANTOINE, Turnhout, 2016 (*Rural History in Europe*, 12), p. 175-191.

BREPOLS PUBLISHERS DOI 10.1484/M.RURHE-EB.5.112267

individual enterprises. Deeply ingrained in this conventional wisdom is the idea that the Second World War was a crucially important catalyst in promoting agricultural specialisation transforming patterns of rural development, ensuring that an increasing proportion of the land was looked after by full time farmers whose main activity was food production. It is the aim of this chapter to critically evaluate these claims.

II. Pre-war regional specialisation

Prior to the outbreak of the Second World War it is commonly accepted that British agriculture was chronically depressed. This manifested itself in low levels of output and productivity, a lack of investment and underutilisation of the land. Under these circumstances, it is tempting to concur with the conventional wisdom that agricultural specialisation, either on a regional or individual basis, would be the exception to the norm. Despite the fact that the 1930s constituted an interventionist period of state assistance for the agricultural sector, heralding the emergence of protectionism, British farmers were unable to compete with low cost imports from the New World and Commonwealth producers. In spite of the formal abandonment of free trade in agricultural produce in 1931, Britain's volume of food imports continued to increase by 6 per cent between 1927-9 and 1938 (Howarth, 1985: 10). By this time more than 90 per cent of wheat originated from abroad and a large quantity of meat was also imported. It was only in the case of perishable foods such as milk that British farmers were able to dominate the domestic market. The only other country which came anywhere near Britain in terms of its dependence on imported food was Switzerland, which remained neutral for the duration of the war. The reasons for these fundamental differences lie outside the scope of this paper but, sufficient to say, the challenges facing Britain in the event of the outbreak of another war were significantly greater than for countries such as Germany which were more self-sufficient in food.

By the 1930s arable farming producing cash crops such as wheat and barley for sale (as opposed to growing tillage crops for animal feedingstuffs) had all but ceased to exist across large areas of central and western Britain. Since the influx of cheap grain in the 1870s, British farmers, unable to cope with low priced overseas competition, had been forced to abandon traditional labour intensive, high cost systems of arable farming, or mixed farming, which had predominated during the period of high farming in the mid-Victorian period.

Regional specialisation was a marked feature of British agriculture. The map (Figure 9.1) needs to be inserted at this point Do you require another copy of the map.? shows, the continuing decline in the area of tillage had resulted in arable cultivation being confined predominantly to the lighter soils of the eastern and southern counties of England by the 1930s. In contrast, arable farming in the Midlands accounted for a relatively small proportion of the agricultural area, and in many western counties it amounted to less than 10 per cent of agricultural land. The existence of such well pronounced regional patterns of land use reflected not only climatic and geographical differences, but also, in Britain, well developed and competitive internal markets for most forms of agricultural commodities, which compelled farmers to concentrate on comparative advantage in terms of the types of crop and livestock they produced.

Figure 9.1. The distribution of arable farming in England and Wales, 1939

Source. Ministry of Agriculture, *Agricultural Statistics, 1939-1945*, (1947), Tables 1-50 in

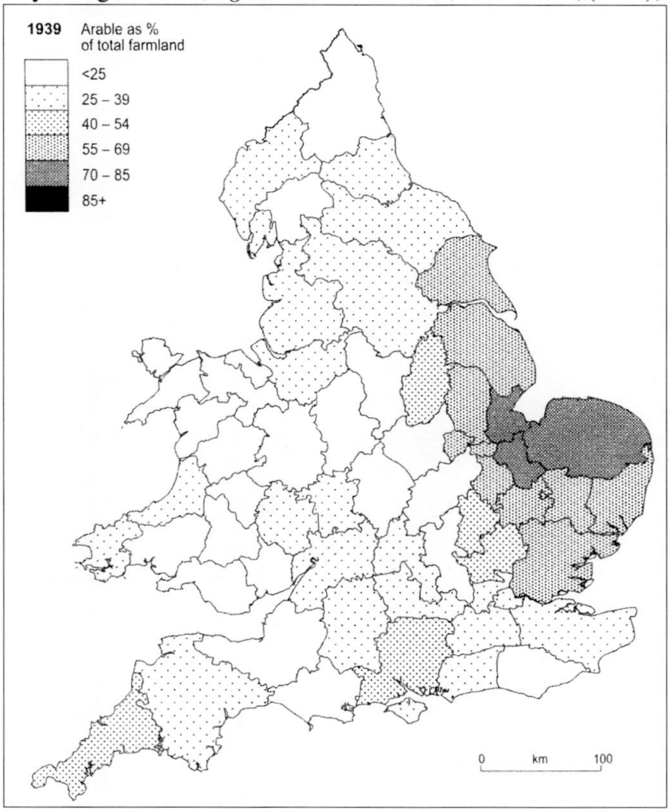

Short, Watkinins & Martin (2007: 5).

British agriculture differed from its European counterparts in terms of its greater dependence on pastoral farming. As Figure 9.2 shows by the spring of 1939 there was an excess of seventeen million acres of pasture land, compared with a mere twelve million acres of land devoted to arable crops, including land left fallow or used for temporary leys. No other European country at this time has such a small proportion of its agricultural land used for cash cropping or such marked regional variations in patterns of land use. Greater specialisation and commercialisation was represented by the variety of farming which prevailed. Vast areas of western Britain were devoted to meat and milk production, with farmers purchasing much of their concentrated feedingstuffs from aboard. Even in the Midland counties of Nottinghamshire, Northamptonshire and Leicestershire where arable farming had flourished during the agricultural revolutions of the late eighteenth and nineteenth centuries, the production of arable crops for sale off the farm was very much the exception to the norm. A disproportionate number of those fields which were ploughed were used to grow fodder crops such as oats and roots for animal feed.

Figure 9.2. Agricultural land use in England and Wales, 1870-1960

Source. Short, Watkins & Martin (2007: 5).

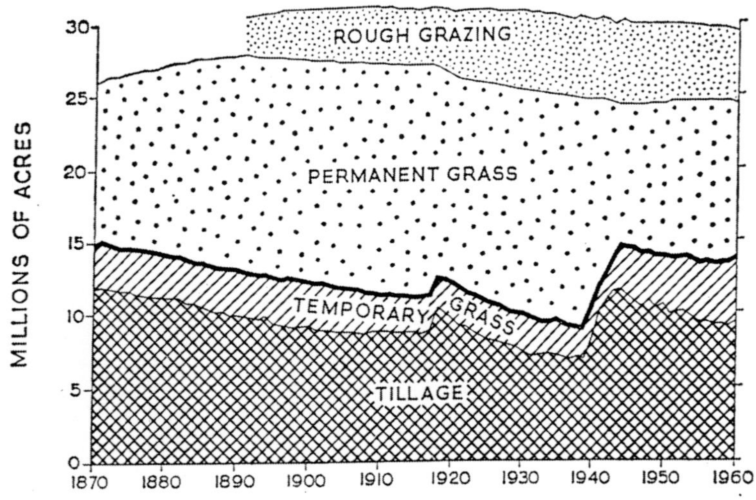

Official accounts have tended to stress the low levels of productivity, measured in terms of output of items sold off the farm, derived from the prevailing mode of pre-war grassland farming. The sector was dominated by extensive low input, low output systems and the grassland improvement was essentially retarded by the availability of low imported feedingstuffs for animals. The most prevalent form of livestock farming, particularly in the western areas of Britain, was that of small scale dairy farming, supplemented by the rearing of beef cattle and sheep on low intensity systems of farm management.

The inter-war deterioration in the standards of farming practice had been compounded by the decline in the restrictions on the way that land could be farmed. Formal or legally binding restrictions were mainly confined to tenant farmers who were required to look after their holdings according to the standards of 'good husbandry' specified in their rental agreements. Following the mass sale of estates which had accompanied the First World War period, there had been a dramatic increase in the number of owner occupiers. Restrictions on the type of crops which could be grown were also relaxed through legislation introduced in the 1923 Agriculture Act.

These changes, coupled with the low levels of agricultural prices prevailing prior to the Second World War, resulted in the countryside assuming an appearance of neglect. By the inter-war depression, vast tracts of lowland England had been allowed to revert back to indigenous grasses of low productivity. Not surprisingly many farmers had ceased to use their grassland as a crop at all but had allowed it to deteriorate to such an extent that it had become little more than an exercise area for livestock.

Historically critics tended to attribute the problems to the lack of entrepreneurial flair exhibited by the farming community. Prior to the Second World War the average farmer had between fifty and two hundred acres at his disposal, too few to be economically viable, and lacked adequate credit and capital. Typically he was constrained by geographical and occupational immobility, originating from a farming family and, according to Calder, shared 'its limited perspectives and its distrust of education, and had tended to ignore the startling advances in agricultural science which were taking place' (Calder, 1969: 483). More recent research has questioned the validity of these interpretations[2].

Specialist producers who concentrated on a single enterprise did exist in this period, particularly in the pig sector where they were able to exploit economies of scale in order to compete with overseas competitors such as Denmark. In the poultry sector a number of large scale producers also characterised the industry.

In terms of land use the most important group was the growing number of specialist arable farmers who had adopted mechanisation in the form of the tractor and newly developed combine harvester. Farmers on the lighter soils in the southern and eastern countries, who ran large units devoted entirely to the production of wheat and barley, were able to compete with overseas imports. In a similar way, large scale, pre-war potato production was confined to naturally fertile areas such as Lincolnshire which were suited by dint of soil type and topography to producing high yields.

---

2 For a detailed analysis of interwar farming see Brassley (2006b).

Extensive, low input, low output systems of mixed farming were dominated increasingly by milk production in the 1930s since it provided a monthly income from the newly established Milk Marketing Board. While a number of specialist dairy farmers such as George Odlum focused their energies on breeding strains of high yielding dairy cattle, their efforts were very much the exception to the norm (Martin, 2007: 237). Instead the vast majority of farmers attempted to survive by economising on all kinds of expenditure, including allowing the quality of their grassland to deteriorate and depended resorted to the feeding of low priced imported cereals Low cost production methods frequently employed the outdoor bail system, where cows were machine milked in a movable bail which was transported from one field to another, with the animals living outside even during the worst periods of inclement weather.

Although this type of production was confined to the lighter soils of the southern counties of England, where the climate and topography was more favourable, it enabled farmers here to compete very effectively with the more conventional and expensive methods of milk production in traditional cowsheds Leading exponents of this new system were the Hosier brothers, who by the late 1930s had more than 25,000 acres devoted to outdoor milk production. Rex Paterson, a self made farmer with an international reputation for his large scale methods of outdoor milk production had, by 1942, more than 10,000 acres in Hampshire and neighbouring counties devoted primarily to this specialist branch of low cost milk production (Martin, 2013, 295-307).

A further unsung pioneer in this field was George Odlum, a world renowned agricultural consultant who, in 1928, had purchased Manor farm, Manningford, Wiltshire where he established a herd of British Friesians. The Manningford prefix which he registered with the Friesian Society rapidly became synonymous with the name of its owner, a nationally acclaimed breeder of pedigree Friesians (Odlum, 1947: 86; Martin, 2007: 236). By the late 1930s he had bred fifty-two cows which produced more than two thousand gallons a year, some of which managed to exceed the three thousand gallons, at a time when the national average was a mere 407 gallons[3]. As a consequence his cows and bulls were widely sought after, and the famed Manningford bloodlines were dispersed throughout Britain as well as other countries. From the onset he was determined to produce a pedigree herd of disease free cows which tested free from tuberculosis. As he explained in his memoirs, 'my conscience would not

[3] The average output per cow in terms of gallons per head per year sold off the farm was 407 for the period 1935-1939. This did not take into acount the amount of milk consumed on the farm. Nevertheless the output per animal was less than 500 gallons. See OJALA (1952: 205).

permit me to produce milk for babies to drink that I would not drink myself' (Odlum, 1947: 3-6; Martin, 2007: 238).

The efforts of these pioneering individuals of the 1930s were in stark contrast to their peers, whose average herd size was little more the twenty cows (Martin, 1992: 216-39). Farmers with so few cows were simply attempting to scratch a living from a multitude of enterprises in which they were neither specialists in terms of the detailed knowledge required to be efficient producers, nor were they large enough to exploit economies of scale.

Given the prevalence of small units of production it is not surprising that farming incomes remained dismally low. According to the National Farm Management Survey, those with holdings of between 200 and 250 acres were making about £250 a year, while the average small farmer with between fifty and one hundred acres was earning on average about £3 per week. Out of these earnings the farmer had to pay his living expenses, taxation, interest on borrowed money, and put aside interest on capital investment (Williams, 1960: 73). Expenditure on non-essential items had to be reduced to little more than self-subsistence levels, not dissimilar from the peasant style farming which historically prevailed in continental Europe.

There were wide variations in profitability between one holding and another, depending not only on size but also on the type of activities pursued and the managerial ability of the family concerned. High levels of self-sufficiency existed on many of the holdings, often exploiting family labour as it was the resource most readily available at minimal cost.

III. Occupational specialisation

Given the economic conditions circumscribed by low prices in the inter-war period, there was little incentive to maximise production in terms of output per acre or holding. Instead the focus was on producing a variety of items, not necessarily for sale off the farm. Many farming families were compelled to seek alternative means of supplementing their income.

Pluriactivity was commonplace, with farmers pursuing a variety of non-farming or complementary activities. According to one contemporary source, there were about 80,000 part-time and spare time farmers whose income was derived primarily from other sources. In addition there were a further 7,500 hobby farmers who either owned or rented in the region of 3 million acres of England and Wales (Astor and Rowntree, 1946: 72). In a survey of Buckinghamshire undertaken by Edgar Thomas it was revealed that only 55.8 per cent of holdings in this area were run by full time farmers. (Thomas, 1938: 21).

After the start of hostilities in 1939, the opportunity for farmers to indulge in non-farming activities was constrained by the fact that full-time agriculture was a reserved occupation. Unlike many other types of rural work, it exempted them from conscription into the Armed Services. Consequently individuals with multiple occupations could be forced to sign up. Even agricultural workers on smaller farms with, according to the local Tribunal, insufficient work to justify retaining them, could be conscripted. Larger farmers and their sons were considerably better placed to be exempted.

IV. Wartime rationalisation and the role of the State

The landmark legislation of the first world war Defence of the Realm Act or 'DORA' was revived under a new name the Emergency Powers(Defence) Act. This empowered the Ministry of Agriculture 'to preserve and maintain agricultural land solely for the production of food, to control, by order the cultivation, management and use of the land in order to secure maximum production from the farms; to terminate any tenancy of agricultural land where it is considered the land is being neglected or badly cultivated' (Martin, 2000: 37). County based War Agricultural Executive Committees (WAECs) were established to put the national directives into practice.

The decentralised system of wartime control was designed to provide an effective, low-cost way of modernising the agricultural sector, in order to ensure that progressive methods of specialised, high input farming were widely adopted. This transformation was seen by K.A.H. Murray, the official historian of the wartime food production campaign, as 'the greatest triumph… [being] the success of the local organisation' exhibiting 'a crusading enthusiasm to bring about a renaissance in British farming', and noting that the key figures in this process were the 'progressive leading tenant farmers and farming landowners on the County and District committees' (Murray, 1954: 339).

Local committees were responsible for evaluating managerial performance, grading farmers into the three categories of A, B and C, relating to perceived levels of output achieved on holdings. Thus 'A' category farmers were deemed to be achieving levels of output in excess of 85 per cent of the holding's potential, 'B' farmers between 70 and 85 per cent, whilst those in the 'C' category were regarded as less efficient and in need of extensive direction. Nationally less than 5 per cent of farmers were placed in the 'C' category.

This system of classification attempted to differentiate between progressive and entrepreneurial farmers and the less productive, conventional individuals who remained entrenched in low-input, low output methods which were deemed to be uncommercialised. In practice A grade farmers were deemed to be the specialist producers who, by dint of the size of their farm or their business acumen, were able to exploit economies of scale in their farming methods. Conversely the implication was that the C grade farmers were those who were reluctant to embrace new specialist methods. But classification was often carried out in a less than objective way, reflecting the bias of committee members (Short *et al.*, 1999: 37). There was considerable reluctance to classify part-time farmers in the A category on the grounds that they were unable to devote their full attention to agricultural activities. Similarly small farmers were underrepresented on the Committees and often downgraded on the assumption that they were technically inefficient.

Wartime grading reinforced the prevailing view that large farmers were considerably more efficient than their smaller counterparts and that arable farmers were more productive that those committed to pastoral farming. As Table 9.1 shows, 90 per of farmers with 700 or more acres were graded A, whereas only 74 per cent of farmers with 5-25 acres and 76 per cent of farmers with 25-100 acres respectively were classified in the top category.

Table 9.1. Relationship between size of farm and proportion of farmers graded A

Acres of Crops and Grass	*Index of Managements*
5-25	74
25-100	76
100-300	78
300-700	84
700 and over	90

Source. HURD, A. (1951: 43).

There were very few instances in which Committees had to resort to formal sanctions such as the dispossession of farmers from their holdings. This has traditionally been interpreted as suggesting that the overwhelming majority of directives issued met with the approval of the farming community and that only a few inefficient farmers challenged them.

V. The expansion of arable farming

The food production campaign of the Second World War was unprecedented in that it initiated a profound structural realignment of British farming in a relatively short period of time. The focal point of wartime policy was a state directed ploughing up campaign which was designed to bring into cultivation an additional 10 per cent of the existing area of grassland in each county in the lowlands each year. Such a strategy was designed to ensure that arable cropping was promoted in every county, irrespective of climatic differences or topography.

The government's initial target to plough up an additional two million acres was achieved by April 1940 despite abnormally cold weather in January and February. In the following year an additional one and a half million acres of pasture were brought back into cultivation, while in the following year farmers were instructed to 'put forward their maximum effort without regard for to the effect on the crops obtainable in 1944 or subsequent years'. The response acclaimed the official history 'was magnificent' (Murray, 1954: 183-4).

The arable area rose from little more than twelve million acres in 1939 to over seventeen million acres by 1944, the peak year of production, dwarfing significantly the achievements of the food production campaign of the First World War. This strategy had a profound effect on rural patterns of development, particularly in respect of the changes which were imposed on the traditional lowland pastoral counties of the East Midlands. As Map 2 shows, by 1944 the proportion of arable land in Leicestershire had risen to over 35 per cent of the available county, while in the neighbouring counties of Nottinghamshire, Derbyshire and Northamptonshire a similar but not quite so pronounced increase was also evident. Conversely the traditional arable counties of the eastern and southern counties of Britain in which arable farming had managed to survive during the interwar depression were considerably less affected by these changes. In Norfolk, for example, the wartime increase in the area of arable land was only in the region of 10 per cent compared with the 300 per cent rise recorded for Leicestershire.

This transformation was not only an economic response to market forces through higher prices, but also the state direction of agriculture under the jurisdiction of the WAECs, which allocated individual farmers with targets in respect of the area of additional land to be brought into cultivation, identifying specific fields to be ploughed up. Farmers were instructed to grow wheat and barley and also potatoes, a crop which the vast majority of pastoral livestock dominated farmers had little previous experience of on a field scale. In effect wartime controls led to a revival of mixed farming, where a multitude of enterprises had to be undertaken in every county and on virtually every holding.

Figure 9.3. The distribution of arable farming in England and Wales, 1944

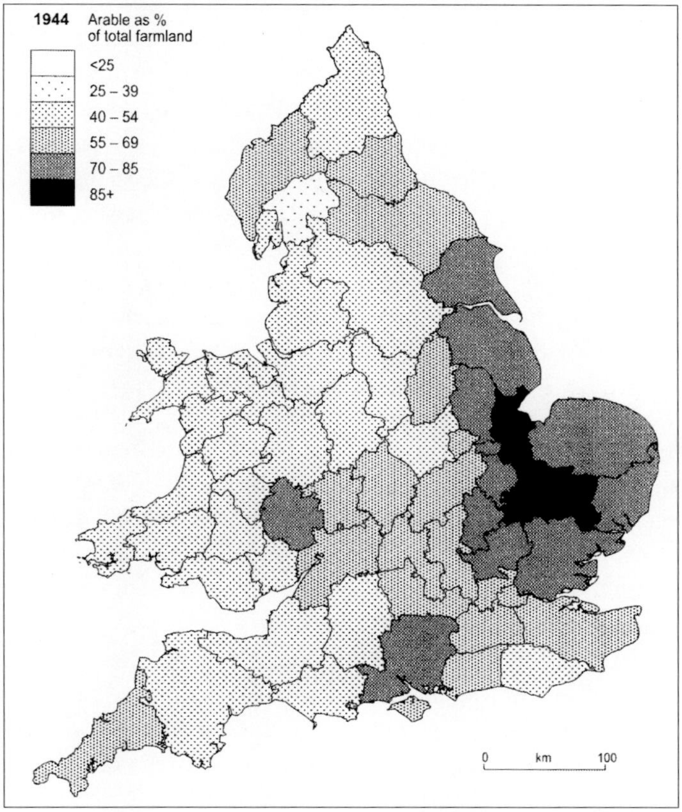

Source. Ministry of Agriculture, *Agricultural Statistics, 1939-1945* (1947), Tables 1-50 in SHORT, WATKININS and MARTIN (2007: 30).

VI. The rationalisation of livestock production

Livestock production was deliberately curtailed by the state in order to allow more land to be used for arable cropping. Compared with pre-war levels, by 1944-45, milk output had declined by 3 per cent, beef and veal production output by 8 per cent, mutton and lamb production by 28 per cent, pigmeat by 65 per cent and egg production by 44 per cent (Murray, 1954: 183-4).

The dairy sector was granted preferential treatment in terms of the allocation of scarce resources such as feedingstuffs. Nevertheless the enforced wartime ploughing up of grassland had a disproportionate effect on large scale dairy farmers such as the Hosier brothers in Wiltshire. As a result of wartime directives they had to bring into cultivation vast tracts of their extensive landholding, at the same time curtailing the size of their dairy herds. A similar fate befell Rex Paterson, a large dairy farmer in Hampshire, who was instructed to convert his farms into arable units. Deemed by the National Farm Survey to be an A grade farmer, the WAEC accused him of adopting a cavalier attitude to their instructions in particular changing the cropping sequence of particular fields and deemed that he should be subject to the stringent controls and regulations imposed on less efficient farmers. He. was instructed to grow more than 600 acres of potatoes, a crop which which was ill suited to the light brashy soils which he farmed and he had no previous experience of. Following a lengthy dispute his local WAEC, a report by the local branch of the National Farmers Union recognised the 'vindictive policy of the Hampshire WAEC, a policy which is responsible for dispossessing quite a large number of farmers' (Martin, 2000: 64).

Ironically, during the Second World War a number of large dairy farmers abandoned milk production completely. The classic example of this was George Odlum, owner of the prestigious Manningford herd of British Friesians, who had popularised the development of progeny testing of dairy cattle on scientific lines. Unable to cope with wartime restrictions on the availability of purchased feedingstuffs he decided voluntarily to sell his farm. His exit from the industry was a major setback for the development of more productive strains of Friesian cattle. His long term legacy was the establishment of the Manningford prefix for pedigree Friesian bloodlines. Although the Manningford herd was dispersed before its genetic potential could be fully realised, Odlum made a lasting impact on improving the breeding of Friesian cattle. His book, *An Analysis of the Manningford Herd of British Friesians*, was originally published in 1943, with the intention initially of distributing it to a few of his friends (Odlum, 1947: iv). But the wisdom and perceptive comments which he had distilled into ensured that the text rapidly became 'one of the best known works on cattle breeding in this country' (Change to Martin, 2007:247).

In addition to the wartime rationalisation of the number of dairy cattle, the shortages in the availability of concentrated feedingstuffs initiated a decline in the amount of milk produced per animal. In order to bolster overall production many small farmers were directed by their local committees to embark upon dairy production irrespective of their previous experience. Often their buildings and land were ill-equipped for clean milk production. The number of registered milk producers increased from less the 133,000 farmers in 1938 to in excess of 159,000 by 1944 (Mingay, 1982: 293). By this time 84 per cent of dairy farmers were small scale producers with herds of less than twenty-five cows. In contrast only 3 per cent of herds had fifty or more cows (which was the number required to justify the use of specialist machinery to enable overheads to be spread over sufficient levels of production, to enable disease control to be managed efficiently, and to allow effective arrangements to be made for hours of work and regular holiday periods for agricultural workers).

A similar scenario was evident in the case of beef cattle and sheep. Rationalisation was undertaken on most agricultural holdings, although production was not completely eliminated. In respect of pig and poultry production, intensive producers who were almost entirely dependent on purchased feedingstuffs were compelled to either drastically curtail production or, in a number of extreme cases, cease production completely. This had a disproportionate effect on technically efficient, specialist holdings which concentrated primarily, if not exclusively, on these types of livestock.

In a desperate attempt to alleviate plummeting pig and poultry production, the Ministry of Agriculture was instrumental in promoting small-scale, inefficient, back yard or non-commercial producers. Tens of thousands of people were encouraged to keep a few hens in their back garden, on allotments, communal areas or on other small plots of land, in order to provide eggs and meat to supplement their meagre wartime diets. Both urban and rural dwellers were encouraged to establish pig clubs to combine their efforts in order to provide enough swill and waste products to keep pigs on a communal basis. The incentive for their efforts was the opportunity to secure a joint of meat when the animals were slaughtered.

While the efficiency of these activities was questionable in an economic sense, they were officially encouraged in order to foster the idea of successful improvisation during an age of enforced adversity. Wartime controls, therefore, were instrumental in reviving the traditional system of small-scale production which predominated in the early part of the nineteenth century and was still evident among peasant producers in Europe. Its revival reflected the temporary accident of enemy blockade rather than constituting a viable long term solution for the modernisation of agriculture.

In spite of the widespread view that there were significant increases in food production during the Second World War, more recent research has challenged the prevailing wisdom. The wartime expansion in arable farming was largely offset by the wartime decline in livestock production. Recent revisionist critiques have suggested that the wartimes increases in agricultural output measured in terms of volume production, monetary value or calorific value were considerably less impressive than commonly believed (Martin, 2000: 47-60; Brassley, 2007: 36-55).

VII. Effects on farmers

Consensus has suggested that farmers derived substantial financial benefits from the wartime period of control. J. K. Bowers acknowledged that 'by any standards farmers had a good war' in light of his calculations that net farm income at constant purchasing power tripled during this period (Bowers and Cheshire, 1983: 61). Even allowing for the fact that this relative increase was inflated by abnormally low levels of income prevailing prior to 1939.

Wartime income gains were not equally distributed across the farming community. Identifying precisely how each group of farmers was affected is rather problematic, given that the majority were not in the habit of compiling detailed financial accounts, nor were they willing to subject their activities to financial scrutiny by outsiders. An insight can be gleaned from the analysis of farm accounts undertaken by provincial agricultural economists, in particular those at the Institute of Agricultural Economics, University of Oxford. These studies reveal that larger farmers who concentrated on a limited number of specialist activities fared considerably better than their smaller counterparts. This was particularly true for farmers in the southern counties who were already geared to the large scale production of arable crops. Conversely smaller farmers in traditional pastoral areas, who had little previous experience of growing arable crops, or who had too few acres to grow arable crops on a viable commercial basis, derived substantially fewer benefits.

As the Minister of Agriculture pointed out in 1944:

> These tens of thousands of small farmers are working as hard as anyone in this country. Many of them are making little more, if as much as the farm labourer, many of them are primarily dairy farmers or farmers whose mainstay of pigs and poultry has gone, farmers who we are compelling in the national interest to grow crops for which their farms may be economically be ill-suited. (*Farmer and Stockbreeder*, 1942: 456).

In a similar vein the Ministry of Agriculture noted:

> One farmer might be called upon to plough up most of his farm or revolutionise his whole method of farming, with possible loss to himself, whilst his neighbour engaged in mixed arable farming continues relatively undisturbed with increased profits. These are the fortunes of war, which it's difficult and often impossible to avoid. (Murray, 1954: 358).

One of the most important consequences of wartime control was its impact on the farming community. A new morality based on economic criteria emerged, transforming the yeoman of Britain the traditional custodians of the countryside into what were essentially the yesmen of Britain, endorsing without question the official orthodoxy of high input farming. Farmers were, in effect, encouraged to become technical specialists, implementing the most economically effective ways of maximising food production, but paying scant regard to environmental considerations or the ethics of food production. An important but isolated exception to this general trend was promulgated by Eve Balfour, whose seminal work, The Living Soil, subsequently came to be regarded as the classic contemporary text of the organic movement (Matthews and Harrison, 2004). However, as the leading authority on the subject, Philip Conford has shown that the war in general was a major setback to organic farming[4].

VIII. Conclusion

The view that the Second World War was a crucial factor leading to the emergence of modern agriculture by the promotion of agricultural specialisation and commercialisation constitutes, at face value, a highly attractive and persuasive interpretation. Throughout the post war period there has been a discernable shift from systems of low input, self sufficiency to exploiting economies of scale in order to maximise output and productivity. The rationale for suggesting that the food production campaign of the Second World War was a major catalyst in this transformation is widely acclaimed to be self evident, given that the main legacy of the war was the 1947 Agriculture Act, which established the principles which have determined the policy of agricultural support throughout the post-war period. However more detailed scrutiny reveals that regional specialisation was actually undermined by the war, and that the concentration of enterprises into large scale specialist units was in some instances reversed. Moreover there were a number of cases which illustrate the way even progressive specialist livestock farmers were adversely treated during the war. A number of the innovative pre-war beacons encouraged by agricultural specialist innovators such as George Odlum, were extinguished as a result of the wartime system of regimented control and regulation. While in general farmers were well remunerated for their work, they were nevertheless regimented and directed under a bureaucratic committee system which demanded conformity and in instances retarded agricultural innovation and product specialisation.

Wartime measures were a pragmatic response to the exigencies of food shortages and should not be eulogised as providing a comprehensive framework or blueprint for a state induced modernisation of agriculture in countries which are struggling to progress along the path of modernity. The experience of the food production campaign during the Second World War merely provides lessons which need to be considered

[4] See for example Conford (2006).

with a fair degree of scepticism. Agricultural specialisation, the widely acclaimed precursor of modernisation, was, even in Britain, primarily a post-war development rather than as commonly inferred initiated by the process of wartime control.

Bibliography

ASTOR, Waldorf (Viscount) & ROWNTREE, Benjamin Seebohm (1946), *Mixed Farming and Muddled Thinking*, London, Macdonald, 143 p.

BRASSLEY, Paul (2006a), 'Wartime Productivity and Innovation', in Brian SHORT, Charles WATKINS, John MARTIN (eds), *The Frontline of Freedom: British Farming in the Second World War, Agricultural History Review*, Supplement Series, p. 36-55.

BOWERS, John, CHESHIRE, Paul (1983), *Agriculture, the Countryside and Land Use,* London-New York, Methuen, 170 p.

BRASSLEY, Paul (2006 b), 'British Farming between the wars', in Paul BRASSLEY, Jeremy BURCHARDT & Lynne THOMPSON (eds), *The English Countryside between the Wars, Regeneration or Decline*, Woodbridge Suffolk, Boydell Press, p.187-199.

CALDER, Angus (1969), *The People's War,* London, J. Cape, 656 p.

CONFORD, Philip (2007), 'The Organic Challenge', in Brian SHORT, Charles WATKINS, John MARTIN (eds), *The Frontline of Freedom: British Farming in the Second World War, Agricultural History Review,* supplementary series 4, 1, p. 67-77.

Farmer and Stockbreeder, 24 March, 1942, p. 456.

MARTIN, John (1992), 'The Campaign for Milk Production', in *The Impact of Government Intervention on Agricultural Output and Productivity in the Second World War,* Unpublished Ph. D. Reading University.

HOWARTH, Richard W. (1985), *Farming for Farmers*, London, Institute of Economic Affairs, 143 p.

MARTIN, John (2000), *The Development of Modern Agriculture: British Farming since 1931,* Basingstoke, Macmillan Press, 236 p.

MARTIN, John (2007), 'George Odlum, the Ministry of Agriculture and 'Farmer Hudson', *Agricultural History Review,* 55, II, p. 229-50.

MARTIN, John (2013), 'Rex Paterson (1903-1978) pioneer of grassland dairy farming and agricultural innovator', in Richard W. HOYLE (ed.), *The Farmer In England 1650-1980,* Farnham, Ashgate, p. 295-324.

MATTHEW, Colin, HARRISON, Brian (2004), *Oxford Dictionary of National Biography*, Oxford-New York, Oxford University Press, 61 vol.

MILWARD, Alan S. (1977), *War Economy and Society,* Londres, Allen Lane, 395 p.

MINGAY, Gordon (1982), *British Friesians: an Epic of Progress,* British Friesian Cattle Society, 311 p.

MURRAY, Keith A. (1954), *Agriculture,* London, HMSO, 422 p.

ODLUM, George (1947 ed. 1), *An Analysis of the Manningford Herd of British Friesians,* Wiltshire Gazette.

OJALA, Eric Mervyn (1952), *Agriculture and Economic Progress (1952),* Oxford, Oxford University Press, 220 p.

SHORT, Brian, WATKINS, Charles, FOOT, William, KINSMAN, Phil (eds) (1999), *The National Farm Survey 1941-43: State Surveillance and the Countryside in England and Wales in the Second World War*, Wallingford, Cabi Publishing, 272 p.

SHORT, Brian, WATKINS, Charles, MARTIN, John (2006) (eds), *The Frontline of Freedom, British Farming in the Second World War, Agricultural History Review,* Suppl. 4.

THOMAS, Edgar (1938), *Farms and Estates of Buckinghamshire,* University of Reading, Agricultural Economics Department, 87 p.

WILLIAMS, Hugh (1960), *Principles for British Agricultural Policy: A study sponsored by the Nuffield Foundation,* London, Oxford University Press, 317 p.

III
The impact of agricultural specialisation on traditional socio-economic systems

10. The balance between subsistence and specialisation in the *huertas* of the Segura: an example from the farms of a large nineteenth-century landowner

María Teresa Pérez Picazo †

I. Introduction

This study attempts to locate new elements that can help us better understand the relation between models of agricultural specialisation and rural development processes. Such an understanding has to be based on a specific notion of specialisation, that of multiple activities within the farm itself. The purpose of this activity is to diversify, that is to produce subsistence crops as part of those destined for the market. This deliberate non-specialisation is, as we shall see, a response to the randomness of external conditions both old and new: climatic restrictions and price fluctuations.

The case selected for the study is taken from the irrigable lands of Eastern Spain[1] between the middle of the nineteenth century and the early decades of the twentieth. This was a period during which capitalism had penetrated more deeply into farming and the rural world (Gallego, 2001: 19-20). In reaction to these changes, farmers began to specialize relatively quickly but without abandoning the traditional Mediterranean peasant habit of ensuring their own supply of cereal crops (Pérez Picazo, 1997: 390-391). This practice probably accounts for the persistence of wheat in the complex crop rotations of the *huertas* of Murcia and Valencia and in the haouz of Marrakech.

The *huertano* farmer adapted relatively smoothly to the new era, probably because of the unique, structural characteristics of Mediterranean farming (Dominguez, 1993: 98-99). Its history and geography were marked from an early date by an outward-looking openness to market forces, which disposed its actors to adapt to the limitations imposed by its climate by making the best of such advantages as its nearness to the sea and population centres. As a result, peasants produced less of the subsistence crops, were forced to sell products such as wine, oil, pottery and silk in

[1] The land adjacent to the rivers that flowed into the Mediterranean.

Agricultural specialisation and rural patterns of development, ed. by Annie Antoine, Turnhout, 2016 (*Rural History in Europe*, 12), p. 195-212.

BREPOLS PUBLISHERS DOI 10.1484/M.RURHE-EB.5.112268

order to purchase cereals. Economic and social factors contributed to the process, notably the widespread presence toward the end of the seventeenth century of small, or medium-sized family-run units, except in some areas where large farms were common (Pérez Picazo, 2000: 117-120).

Starting in the mid-nineteenth century, economic and political changes increased the use of cash in rural areas, as country folk now needed ever-increasing amounts of ready money to pay for new taxes, land leases and purchases that they could not find on the spot. Never the less, on the irrigated farms, the traditional pattern of specialisation remained unchanged. Crops were diversified but farms still continued to set aside part of the specialized crops – fruits and vegetables, at this point – for subsistence. This was because peasants had to adapt to markets that were still imperfect (Dominguez, 1996: 234-236), reacting to continual crises by resorting to on-farm consumption. The outcome was a solid productive organization enhanced by the practices of intensive agriculture and habitual market orientation, but with a sort of 'life insurance' built in. This probably explains why the system lasted well into the 1950s and 1960s and why the descendants of the old leaseholders hung on so long to the farms they had worked, as they said, from 'time immemorial'.

II. The *huerta* of Murcia: basic physical, socio-economic and institutional features.

To describe the functioning and evolution of the specialisation model defined above, we have taken the case of the irrigable land (*huerta*) of Murcia[2]. This land was characterized by a singular agricultural and economic system (Pérez Picazo & Lemeunier, 2003: 175-177). During the nineteenth century, total farm area in the *huerta* ranged between 10,992 and 12,048 hectares. The climate was typical of south-eastern Spain, with a high annual average temperature (17° C) and low, variable rainfall (360 mm.). The shortage of water and irregular levels in the Segura River, on which local irrigation depends, further marked the unique character of the area, accounting for the random nature of crops and the early entry of huertano farmers into the market, as described above.

Despite these difficulties, there is no doubt that the irrigable land of the Segura, particularly the huerta around the regional capital, represents the most fertile portion of the Murcia region, where population was the most dense. Figure 10.1 shows the population change in relation to the region's UAA (Utilised Agricultural Area). As with monsoon countries, it cannot be said that this density alone would have pushed

---

2 The *huerta* of Murcia lies around the city of the same name. Murcia is the capital of a region watered by the Segura and bordering on the Mediterranean.

the nineteenth-century population of the *huerta* into a demographic transition, which is one of the most reliable indicators of social and economic modernisation. On the contrary, research carried out by several specialists has demonstrated that this population density was itself part of a high-pressure demographic system that was typical of the Old Regime, and characterised by very high birth rates needed to make up for an equally high mortality rate (Martínez Carrión, 1988: 92-93). High mortality rates persisted until relatively late, into the early decades of the twentieth century, thereby confirming the traditional nature of the population model.

Figure 10.1 Map of Murcia

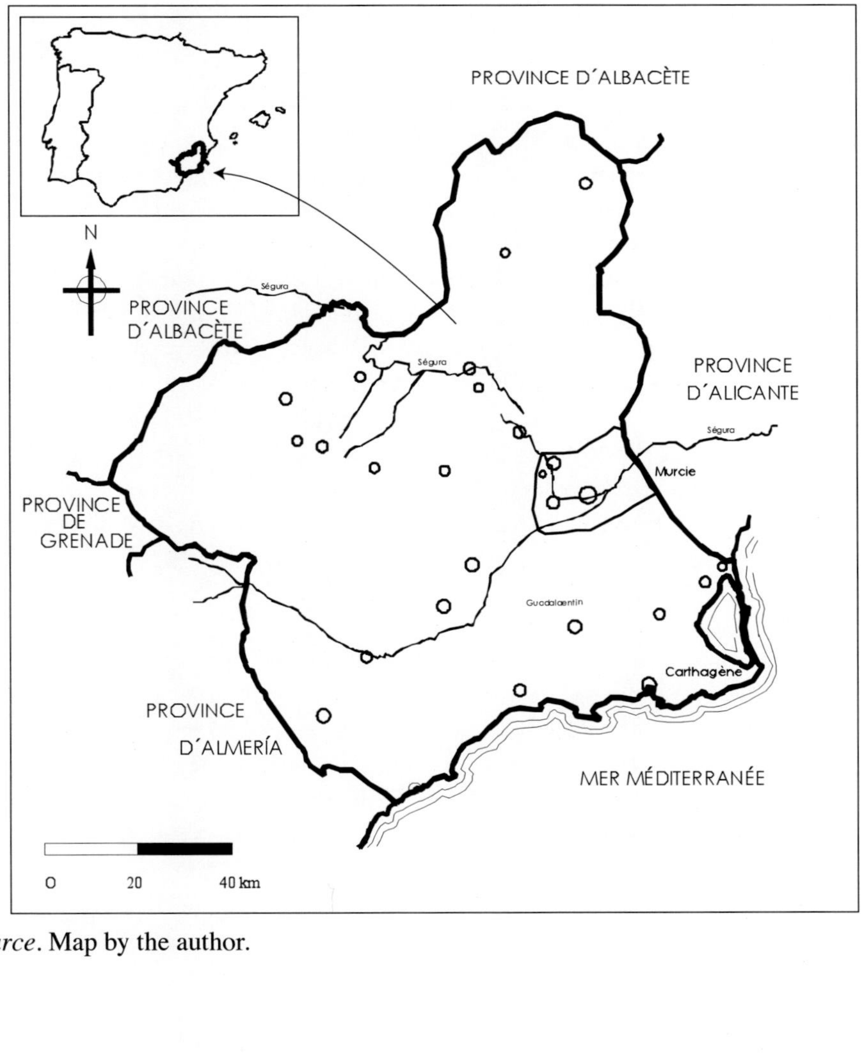

Source. Map by the author.

Table 10.1. Development of irrigated land and demographic density in the *huerta* of Murcia, 1756-1920

Year	*Area (hectares)*	*Population*	*Density (inhab./UAA)*
1756	10,201	25,220	125
1836	10,425	52,144	256
1857	10,992	62,426	304
1879	10,747	64,770	325
1920	12,408	98,822	496

Source. 1756, Catastro de Ensenada [Ensenada's Cadastre]; 1836, 1850, 1879 and 1920, Padrones de Tahúllas de la Junta de Hacendados [Board of Landowners' Census of Tahúlllas].

Over time, irrigated land tends to alternate between dynamic periods or cycles and times of stagnation or even decline. The dynamic periods normally coincide with the successive appearance of vector products, that is, outputs showing a certain degree of growth and that stimulate the investment needed to expand the area of land under irrigation. This is what happened with silk during the sixteenth century and horticulture in the early middle decades of the nineteenth (Pérez Picazo, 2004, 12-14). Variations on this pattern occurred between 1756 and 1836 and from 1836 to 1857, when new crops began to spread, but progress was interrupted between 1857 and 1879 when new irrigation became more infrequent and irregular, as plots of land farther away from the Segura were brought under cultivation (Table 10.1). At this point, a technological 'bottleneck' occurred that was not overcome until large-scale hydraulic control systems were introduced in the last two decades of the nineteenth century and the early decades of the twentieth. At that point, more use was made of underground water and storage reservoirs, resulting in a rise in irrigated hectares between 1879 and 1920 (Calatayud & Martínez Carrión, 1999: 16-17).

These developments show how decisive the new cycle of agricultural specialisation was for the modernisation and progress of irrigation farming. As the crops it made possible were highly profitable, it became a powerful lever making it easier for large-scale hydraulic systems to take the place of small- and medium-sized ones. We should note that fruit and vegetable crops followed their own distinct paths. In an attempt to strike a careful balance between subsistence and marketable products, farmers inserted vegetable crops into existing rotations. This did not require any significant investment. Fruit production was different. Fruit trees were planted on the edges of irrigable land or on adjacent hillsides. As a result, more land was cultivated and so more hydraulic systems were used to irrigate it. This was a typical example of the

Mediterranean sequence of terrace construction, by which water was pumped up to the terraces, first by water-wheels, and then by electric motors. In its early stages this process required heavy investment (Pons, 1993: 40-42) and was only within the reach of landowners. From about 1880-1890, however, farmers began to take part in the changes by forming small companies.

The physiognomy of agricultural structures deserves some mention here (Garrabou, 2001: 216-218). During the period of over a century from start to finish, their most visible and constant feature was the co-existence of a high level of property concentration (the Gini index rose from 0.82 to 0.84 between 1800 and 1920) and an equally high fragmentation of operations (Pérez Picazo *et al.*, 1992: 163-164). This shows up in the predominance of indirect management, which took the form of small or medium-sized crop units under precarious contracts. The evidence shows that in 1822, 81.4% of the irrigated land was managed in this fashion; a century later, the share had risen to 86.7%[3]. Landowners, conscious of the rising demand for horticultural products and the growth of markets, chose to promote agricultural change, trying to make leaseholders share the risks by offering them the best possible lease terms (Pérez Picazo, 2005: 33-35).

Table 10.2. Appraisal of a *barraca*, 1867

6 stone and plaster posts	300 Spanish reales
6 wooden walls	30
5 wooden interior walls	30
5 chains	35
41 tins	123
40 bundles of cane	40
40 bundles of carex	60
Nails	13
Esparto grass rope and cord	30
Two doors	70
TOTAL	731 Spanish reales

Source. Inventario Post Mortem. APHM. Escribano (notary) Narciso Sánchez, Protocolo 4887, p. 234.

Thus the typical nineteenth-century Murcian *huertano* was, like his Valencian counterpart, neither a day-labourer nor a small landowner, but rather a leaseholder,

[3] Archivo Municipal de Murcia (amm) [Murcia Municipal Archives]: *Padrones de Riqueza Territorial de la Huerta de Murcia* [Territorial Wealth Inquiry for the *Huerta* of Murcia]

who would normally work smaller-sized crop units and probably built a modest home, a *barraca* (a sort of cabin constructed from materials available on the spot). These practices, plus a landscape of scattered hamlets and small villages, generated a nebulous population structure (Millán, 1991: 24-26). The origins of this phenomenon can be traced to the demands made on irrigation (access to water was sometimes made available at late hours of the night) and to the need to supervise land cultivated in fruits and vegetables through the greater part of the year.

Table 10.3. Size of the farms in the *huerta* of Murcia, 1822-1824 and 1922

a) 1822-1824

Frequencies (in hectares)	*Area*	*Number*	*Total Area (%)*	*Percentage of Farms*	*Average Area*
0 to 0.49	485.96	1,767	4.29	29.13	0.28
0.5 to 0.99	994.84	1,392	8.77	22.95	0.71
1 to 1.49	1,104.79	917	9.74	15.12	1.20
1.5 to 1.99	806.99	473	7.12	7.80	1.71
2 to 2.99	1,475.27	607	13.01	10.01	2.43
3 to 3.99	969.19	284	8.55	4.68	3.41
4 to 4.99	842.08	191	7.43	3.15	4.41
5 to 9.99	2,277.00	395	20.09	5.52	6.80
> than 10	2,382.00	99	21.01	1.63	24.06
TOTAL	11,339.03	6,065			1.87

b) 1922

Frequencies (in hectares	*Area*	*Number*	*Total Area (in %)*	*Percentage of Farms*	*Average Area*
0 to 0.49	973.1	3,312	10.48	37.80	0.29
0.5 to 0.99	1,825.74	2,585	19.66	29.50	0.71
1 to 1.49	1,600.00	1,346	17.23	15.36	1.19
1.5 to 1.99	818.95	484	8.52	5.52	1.69
2 to 2.99	1,474.00	628	15.87	7.17	2.35
3 to 3.99	609.42	180	6.56	2.05	3.39
4 to 4.99	365.81	83	3.94	0.95	4.41
5 to 5.99	690.70	105	7.44	1.20	6.58
< than 10	930.51	40	10.02	0.46	23.26
TOTAL	9.288.00	8,763			1.06

Sources. amm. For 1822-1824, Padrón de Riqueza Territorial de la Huerta de Murcia and for 1922, Padrón de Tahúllas from said year.

As can be seen from Table 10.3, farms tended to shrink over the course of the nineteenth century; the average size decreased from 1.87 hectares in 1822 to 1.06 hectares one century later. More important, the proportion of smaller farms (as defined by both their total area and the number of farms in that category) grew, while the share of farms in the larger categories decreased. This was the result of both population growth and the landowners' strategy: they preferred to subdivide farms in order to increase their profits; the figures in the accounts consulted leave no doubt about this (Carmona-Simpson, 2003: 237-238).

One last point to highlight in this process is the role of the landowners and leaseholders. Did the former participate by guiding crop changes and bearing their cost or did they behave like the hated 'landlords' portrayed in the old clichés? Did the landlords actively participate, not only with their labour but also by financing the changeover out of their savings? This research takes us farther towards answering these questions.

III. Presentation of the case study and its sources

The bulk of the sources used for this paper consist of the private accounts of a wealthy family of landowners, the Counts (Condes) of Valle de San Juan, who appear at the top of the Real Estate Tax rolls throughout the nineteenth century[4]. They were descendants of the urban nobility established in the Kingdom of Murcia during the Early Modern period. Most of them had managed to pull together considerable estates by systematic acquiring and splitting up underpopulated land. Primogeniture allowed them to keep estates in the family line. According to a post-mortem inventory drawn up in 1901, the family owned 9,153 hectares of land, 201 urban properties and five mills[5]. The two basic characteristics of this enormous landed fortune were the dispersion of the plots, scattered across parts of eight different municipalities, and a wide range of profitability. The 'crown jewels' consisted of the irrigated lands, representing only 10.5% of the total area but providing 71.36% of the total income. This study concentrates on a part of these latter holdings, the farms located in the *huerta* of the regional capital.

These holdings consisted of a series of plots, not all of them contiguous, amounting to a total of 657 hectares at the middle of the nineteenth century. Sources, to be discussed

[4] Fonds of the Conde del Valle de San Juan (acvsj). Originally kept at the Casa Solariega in Calasparra, and now in the Municipal Archives of that town. The collection contains 56 individual files on the Murcia estates (13 in Caravaca and 5 in Calasparra), running from the fifteenth to the twentieth century.

[5] Archivo Histórico Provincial de Murcia (ahPm) [Murcia Provincial Historical Archives]. Inventory of José Tomás Melgarejo y Musso, Escribano Juan de la Cierva, L. 10032.

below, that identify and record new plots show that this figure was a little lower in 1800 and a little higher in 1902. But the most important change was in the number of crop units: at the end of the eighteenth century, in 1790, there were 310 of them, but in 1860 the number had already grown to 359, and by 1902 there were 422. Concurrently, average farm size had been gradually decreasing, from 2.11 hectares to 1.8 and then to 1.27 hectares. This change was similar to that experienced in the huerta as a whole, but the average figures are slightly higher (Pérez Picazo, 1991: 59-60).

Table 10.4. Extract of the model used for the leaseholding contracts

Introduction: Date, Name and residence of the parties, Location and size of the plot, period and price
Most important clauses: Care of the water channels and payment to the Council of the corresponding sums for the annual cleaning thereof (this sum was referred to in Spanish as mondas)
Cultivation practices in the manner and custom of a good labourer: hours and amounts of irrigation; number of labourers, weeding, etc.
Special care of the trees: if a tree is lost, the leaseholder shall be responsible for its replacement and the cost thereof
Required characteristics of the live capital: one or two shooting animals, depending on the size of the crop operation of a small plot of alfalfa; goats and sheep are prohibited; limited farmyard animals
The construction and care of the *barraca* corresponding to the leaseholder. Upon departure, an agreement shall be made with the entering party regarding the payment of the costs thereof
Specification of the amount of the adeala
The contract is at risk and venture of the leaseholder. In other words, the entire sum of the rent shall be paid even in the case of a total loss of the harvest for any unforeseen act of force majeure
The landowner shall be responsible for the Real Estate Tax
The landowner may introduce changes to the type of crop (trees included) if he takes responsibility for the costs and reaches an agreement with the leaseholder. The latter, on the other hand, must request authorisation from the landowner to do the same
The contract is renewable every four years, the moment at which the landowner may increase the rent, if so required.

It should be emphasized that the accounting records used for this study relate exclusively to farms managed under lease. There were no intermediate forms of lease in the eastern huertas, whether in the Segura (Murcia and Orihuela) or the Turia and the Jucar (Valencia) or in the Mijares (Castellon): that is there were no payments in kind save for the annual Christmas gift to the landowner, termed adeala, normally consisting of a pair of chickens, a turkey or a ham. The leaseholder worked the land and paid a fixed sum of money on the feast of St. John (24 June), but he kept the product of his labour, from which he had to feed his family and earn enough money to pay the rent. Landlords made sure that the rent would be paid by choosing 'hard-working and respectable' candidates. The process of selection was made easier by the Count's use of his 'information capital': his family's social roots in the area provided them in effect with not only knowledge of the social environment but also free access to informants such as local clergy, town mayors, former employees, tradesmen, and the like.

What sort of document makes it possible to study the appearance and development of this collection of farms, and what sort of information does it provide? For quantitative purposes, there were two outstanding sources: the annual balance sheets presented to the landowner by a representative of the general administrator (the *Mayordomo de Campo y Huerta*) and the so-called leaseholder books.

The annual balance sheets cover almost the entire period. Their income section is the only place where the annual rent payments are to be found, along with late payment fines (between 80 and 85 per cent of the leaseholders incurred these fines at some point in their lives), and reimbursements. On the expense side of the sheets were the payments made by the landowner, covering state tax, the wholesale purchase of mineral fertilisers, the execution of irrigation work and fruit tree planting. The difference between the two entries was the net product obtained by the owner of the land. Obviously, this source cannot provide us with information on production and productivity. However, it can tell us about the profitability of the land, which we have reconstructed for the entire nineteenth century, and also about the progress of the region as a whole. We can consequently estimate the landowner's investment strategy regarding both the invested capital and its use.

The second source consists of thirteen volumes of leaseholder books. Information in these is much more abundant than in the balance sheets. They were renewed every five to six years and, although not all of them survive, they constitute a basic source of information. In these books, the base unit is not the leaseholder, but rather the plot of land. Each plot is identified by its geographical location, the irrigation conduit (*acequia*) providing it with its water, its size (with successive subdivisions), the name of its direct farmer(s) and the type of crop. This last piece of information is essential because any change in it automatically implied a rent increase (measured in money per unit of land) or, if it involved new tree planting verified by the landowner, a rent reduction that ran until the trees began to bear fruit. The information makes it possible to do a survey of all the farms maintained by the Casa Condal throughout the entire period, taking into account sales and purchases; as a result we can carry out a proper case study. Above all, it allows us to measure the impact of the landowners' and leaseholders' initiatives and methods on the development of specialisation and intensification over time.

Thirdly, there also exists a wide variety of other documentation. The most important are the Count's instructions to his employees on how to carry out farm inspections, both by looking them over and by interviewing the people involved. The main aim of these inspections was to supervise the introduction of new crops, the care of the trees and the state of the irrigation channels. Sometimes it was done by a simple method of posing short and concise questions, requiring similar short and concise answers which would then form the basis of general conclusions. This was standard practice during the time of the fifth Count, José Tomás Melgarejo y Musso, who, more often than not, used this information to make his decisions. It is also worth mentioning the copies of the protocol act books (*libros de actas de protocolos*), almost always related to purchases and sales, contracts, wills and inventories, both of the family and of various leaseholders who had asked the Count to act as their executor. Finally, there is of course, a large quantity of correspondence, both of the family and on matters related to the farms. Among this correspondence, it is worth highlighting the *Libro de mi trabajo*, a diary written by the above-mentioned José Tomás Melgarejo between 1865 and 1890, which has enabled us to obtain first-hand the point of view of a landowner during a period of change.

This collection of documents sheds useful light on very diverse aspects of estate management and the daily life of the farmers. However, the quality and quantity of the papers conserved depended on the personality and priorities of the successive owners of the estate. The third owner, Concepción González de Avellaneda, ran the estate administration very carefully but also cautiously. When he died young without a direct heir, he was succeeded by two very different characters, José Maria Melgarejo y González de Avellanada (1812-1862), fourth Count of Valle de San Juan, and José Tomás Melgarejo y Musso (1839-1902), the fifth Count. The fourth Count, an active liberal conspirator, spent part of his life in exile in France and played no role whatsoever in his estates during the final years of his life. By contrast, the fifth Count, Melgarejo y Musso, took great care of the estates and improved them, investing significant sums of money in their modernisation. He even spoke German and travelled to Germany to gain further knowledge of agricultural progress; his library contained a study by Liebig.

IV. The role of the landowner: cacique or agricultural businessman?

Why did rent-dominated agricultural systems persist? In one of his letters[6], the fourth Conde del Valle summarises what must most of his peers must have thought. Firstly, he states that the lease-holding system is used for labour-intensive crops; direct management of day-labourers on these would have raised production costs. Furthermore, he adds that the widespread dispersion of the plots in the interior would make it hard to lease them out in large blocs. The Count therefore judged that the existing system was the most suitable for developing intensive and specialised agriculture with the least possible amount of risks. In modern economic terms, the choice was logical, given the nature of the market factors: strong demographic pressure and property concentration favoured imposing and intensifying the use of labour rather than of land.

Rent became the main form of rural surplus extraction and capital accumulation, as occurred on other *huertas* (Calatayud-Millán, 2000: 79-81). In order to update the information contained in the old documents, agricultural surveyors measured the plots between 1830 and 1840, and crop supervision became constant. Attempts were made to ensure that the direct farmer did not benefit to a greater extent than the landowner. As a result, rents were raised almost everywhere around 1840, in the 1860s and in the period from 1890 to 1900. The first set of increases was made possible by the changeover from pepper to paprika; the second by the rise in wheat prices[7]; and the third by the rapid expansion of horticulture as a direct result of measures taken by the landowner.

Despite fluctuations, the upward trend in profitability during the nineteenth century shows that the aims were satisfactorily met (Pérez Picazo, 2005: 48-49) (Figure 10.2). True, there was a drastic decline during the first two decades of the century, resulting from the impact of the War of Independence and the political conflicts between absolutists and liberals. But then followed a long period, from 1830 to 1880, during which the general trend continued upwards. Such fluctuations as existed can be explained by the bad administration of the fourth Count, the silkworm plague and, as indicated earlier, the slowdown of horticultural expansion. Even the great depression of the late nineteenth century had only a moderate effect: falling wheat prices were made up for by the recovery of horticulture (Garrrabou, 2001: 229).

---

6 ACVSJ, José María Melgarejo González de Avellanada to the Marqués de Ordoño, Murcia, September 5, 1851, Fondos de Murcia, L. XXX.

7 ACVSJ, Fondos de Murcia, l. xxv, Administrator to José Tomás Melgarejo y Musso, April 3,1850, suggesting a rent increase: 'When the plots were leased, a *fanega* of wheat was worth 36 reales, so the land surveyors set a price of 56 reales for a first-rate *tahúlla* (1 *tahúlla* = 0.1118 ha). But, noting that prices had increased by 50%..., they agreed to raise the price accordingly'.

Figure 10.2. Net landed profits, 1800-1900
(Five-years average and constant prices from 1800)

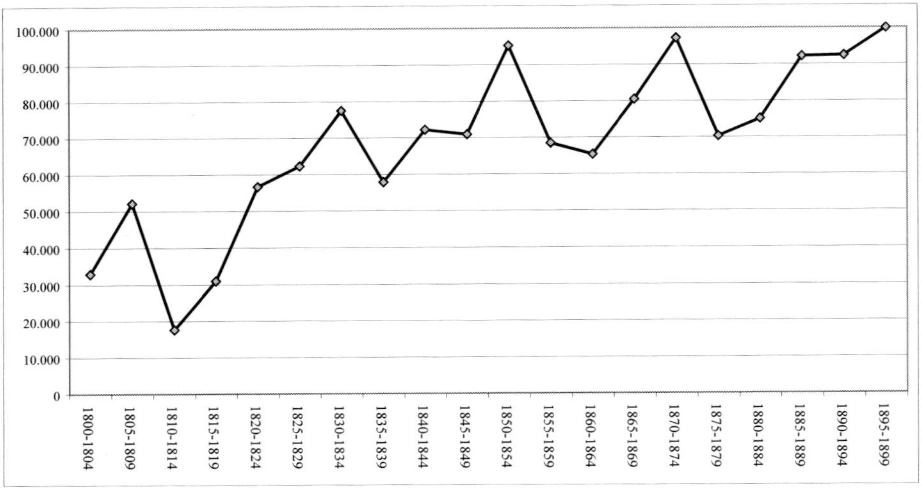

Source. acvs. Bilans 1800-1900. Fondos de Murcia, l. x-xv. Deflators, price of wheat and silk.

In the light all this, we can assert that the Condes del Valle behaved as agricultural businessmen and that their basic priority was to make a profit. They certainly succeeded at doing it: at the end of the nineteenth century, the estate had grown by 765 hectares (715 of dry land and 50 of irrigated land) and farming had been modernised. On the *huerta* of Murcia alone, 189 hectares of fruit trees – citrus fruits, apricots and peaches – had been planted, representing 28 per cent of the expansion; and irrigation levels had been raised by improving the irrigation network (*acequias*) and installing two waterwheels and three electric motors. Total investment reached 636,891 Spanish reales, 345,578 of which went to agricultural improvements and 291,313 to land purchases. The high figures for improvements came from the costly planting carried out, mainly on the mountainsides surrounding the huerta. Drainage was better there than on the valley floor but it was necessary to level the slopes, uproot the old mulberry trees, bring up water from the river and wait an average of 8 to 10 years for the new trees to give fruit. Moreover, these sums were almost entirely paid for by the fifth Count during his nearly forty-year 'reign'; the only exception being the 41,286 Spanish reales laid out by the fourth Count, José María Melgarejo Salafranca, for repairs to the *acequias*.

The resulting relationship with the leaseholders was shot through with a paternal authoritarianism that is difficult to describe. The Condes allowed up to three consecutive periods for the payment of overdue rent, occasionally paid for medical services, acted as executors of wills and defended the irrigator in lawsuits before the *Consejo de Hombres Buenos*, the highest authority of the huerta. However, they would not tolerate the slightest disobedience, bad manners or attempts to rebel. Any such aberration was punished by non-renewal of the contract. Further control over the farmers in general was assured by confiding the positions of town mayor, or membership in the *Consejo de Hombres Buenos* to trusty men, and even by pressuring the Bishop of Murcia to name reliable parish priests. Above all, it was obligatory to vote at elections in accordance with the wishes of the owner of the Casa. The abundant correspondence to which we have had access recounts many examples of such conduct, which, in our opinion, places the last two Condes del Valle among the group of *caciques* described as 'good' by the standards of their time, but tyrants nonetheless (Millán & Zurita, 1998: 153-154).

V. The role of the direct farmer. Mixed farming, dependency, and deteriorating quality of life

What was the real situation of the leaseholder community? It could surely be described in terms of dependency, self-exploitation and mixed farming.

Dependency, firstly. Because of the configuration of market factors, the modest farmer was in a very fragile situation vis-à-vis his all-powerful tyrant of a landowner However, dependency did not apply only to relationships with the landowner. In actual fact, products like silk cocoons or pepper for paprika were sold through middlemen, the *corredores* and the wholesale *Casas de Comercio* established in the regional capital. The lack of information, the urgent need to sell the harvest rapidly when prices were at their lowest in order to pay the rent, the 'advance' loans granted by the above-mentioned establishments and petty consumer debts incurred with local shops, often placed the farmers in a very difficult position.

As we have seen, the size of the crop unit gradually decreased while rent increased. In 1840, a leaseholder had to pay between 56 and 85 Spanish reales for a first rate *tahúlla* (equivalent to 1,118m^2). In 1860, this rent had risen to between 112 and 163 and in 1900 it cost between 153 and 298[8]. The cost was met by self-exploitation and the use of mixed farming.

Self-exploitation, in the short term, took the form of cutting back on family consumption in order to pass on a larger part of the surplus to the market. However,

8 ACVSJ, Libros de arrendamientos, [Leaseholder Books] Fondos de Murcia LXXVI and XXVII.

before long a different solution was found: introducing other marketable products into crop rotations. This increased income but also the work required, when the fallow season had been eliminated. The choice of this solution resulted from a series of circumstances: the need to survive; the economic possibilities opened up by new land uses when they coincided with increased demand, a strongly polarised social structure, demographic pressure and so forth. It should not be forgotten that, as Chayanov claimed, the important thing for the modest farmer is the final outcome, that is, physical and economic reproduction, and not the work required to achieve it.

That work was carried out in a carefully conceived strategy in which the whole family took part. The women, with the help of the children, bred the silkworms; they travelled to town markets to sell fruit or poultry produce and they looked after the home. The men found work as day-labourers in the heavy work of agriculture. In the *Libro de mi trabajo* written by the fifth Count and in his correspondence with administrators and friends, numerous examples of this pattern of behaviour can be found, especially on farms less than a single hectare in size. The Padrones de Vecinos de la *Huerta de Murcia*, which list the professions of all people over 13 or 14 years of age, are a good source of information on this.

We lack direct testimony on the appearance and operation of the farms,. However, there is an alternative: the post-mortem inventories of leaseholders who appointed the successive Counts as executors of their wills, and which were included in the registers of notarized acts (*libros de Actas Notariales*) deposited in the archives[9]. A sample has been put together of such inventories from 30 small and medium- sized farms (between 0.1 and 5 hectares), dating from the years during which specialisation accelerated.

Table 10.5. ***Post mortem*** **inventories of the leaseholders, 1849-1890**
(farms from 0.1 to 5 hectares)

Categories (%)	*Crops* (%)	*House-hold* (%)	*Live-stock* (%)	*Reserves* (%)	*Farm Equipment* (%)	*Debts* (%)	*Ready Money* (%)	*Invent. Value* (rs = reales)
0.1-1	51.7	6.2	7.9	7.2	9.8	16.6	0.6	500-1,000 rs
1.1-2	44.1	8.3	13.4	10.3	12.1	10.1	1.7	1,000-3,000 rs
2.1-5	9.7	14.7	16.1	12.7	15.4	7.1	4.3	3,000-6,000 rs

Sources. ACVJS, L. XXVII.

[9] ACVSJ. For 1849-1870, Escribano Deogracias Serrano, Fondos de Murcia, LXVIII; for 1871-1890, Escribano Miguel Cano Cordero, Fondos de Murcia, LXXX.

There are two notable findings. First, the high percentage of the value of inventories given over to labour, seed, stock and crops, especially on crop units of less than 2 hectares, and the very low percentage of total value accounted for by the items of everyday life in categories such as household goods (clothes and furniture) and agricultural activities (livestock and farm equipment). These figures paint a picture of a very hard life: a fragile *barraca* for a home that hardly has space for furniture, little livestock (only the draft and farmyard animals) and meagre reserves. Secondly, these meticulous listings indicate that farm equipment corresponded to traditional irrigation tools and that the turn-plough was not yet widespread. With this lack of resources, the life of a farmer was surely an exhausting one.

VI. Conclusions

Two features stand out from the transformations described above: a unique model of agricultural specialisation and the vitality of a rural society that succeeded in adapting both to circumstances and to its environmental limitations (González de Molina, 2001: 66-67). There can be no doubt that this system effectively monetised the economy and consolidated agricultural capitalism. Increasingly high rents led leaseholders to specialise, while forcing them to increase their marketable surplus and to look for seasonal jobs as a day labourers. There is equally no doubt about the fundamental role played by the *huerta* community. Peoples' ingenuity helped them find ways to search out new sources of ready money and to develop new crops without abandoning those needed for consumption on the farm.

It is worth recalling here that this phenomenon is in no way peculiar to the area under consideration. In fact, there are numerous similar cases in other Mediterranean regions. The vineyards of Catalonia and Languedoc-Roussillon and certain Greek regions such as Thessalia and Corinthia, are examples of other such cases. In all of them, the small family-run farm, whether owned or leased, was both more adaptable and more resistant to the forces of market orientation than larger farms managed directly or worked by salaried labour. Könning's intelligent analysis (Könning,1994: 108-109) goes a long way to explaining this, in its suggestion that the majority of peasant farms found a subsistence opportunity in specialising their production, and were even consolidated by it.

The system, though efficient from the standpoint of capital accumulation, was not so from the perspective of human capital. In most cases, the success was owed to a significant increase in workload and some deterioration in the quality of life. This is not an exceptional situation: other available studies also confirm that this decline is an characteristic of the first stages of capitalism. Using an anthropometric measure,

Martínez Carrión showed that the height of farmers on the *huerta* of Murcia declined slightly between 1840 and 1890 (Martinez Carrión, 1986: 82-83). Economically, this process also reduced the education level of human capital. This had a decisive influence on the progress of labour productivity. This is a very difficult variable to calculate, but our data on agricultural technology and the quantity of farm livestock seem to show a stagnation in these figures.

Matters are clearly different with regard to capital accumulation (Robledo and López, 2007: 26-30). The figures for the increase in landowner estates and the level of investment produce a clearly positive balance. The context was favourable: despite the spendthrift fourth Count, his successor's relatively moderate efforts to modernise not only recovered what had been lost but also converted the family estate into an administrative model praised by the local press.

Bibliography

CALATAYUD, Salvador, MARTÍNEZ CARRIÓN, José Miguel (1999), 'El cambio técnico en los sistemas de captación e impulsión de las aguas subterráneas para riego en la España mediterránea', in R. GARRABOU, J. M. NAREDO (eds), *El agua en los sistemas agrarios*, Madrid, Argentaria / Visor, p. 79-107.

CALATAYUD, Salvador, MILLÁN, Jesús *et al.* (2000), 'El rentismo nobiliario en la agricultura valenciana del siglo 19th', *Revista de Historia Económica*, XVIII, 1, p. 79-107.

CARMONA, Juan, Simpson, James (2003), El laberinto de la agricultura española, Zaragoza, Prensas Universitarias Zaragoza, 366 p.

DOMÍNGUEZ, Rafael (1993), 'Caracterizando al campesinado y a la economía campesina. Pluriactividad y dependencia del mercado como nuevos atributos de la campesinidad', *Agricultura y Sociedad,* 66, p. 97-136.

DOMÍNGUEZ, Rafael (1996), *El Campesino adaptativo. Campesinado y mercado en el Norte de España, 1750-1880,* Santander, Universidad de Cantabria, 256 p.

GALLEGO, Domingo (2001), 'Sociedad, naturaleza y mercado: un análisis regional de los condicionantes de la producción agraria española, 1880-1936', *Historia Agraria,* 24, p. 11-57.

GARRABOU, Ramón (2001), 'Crecimiento agrario, atraso y marco institucional', in Josep PUJOL *et al.* (eds), *El pozo de todos los males,* Barcelona, Crítica, 2001, p. 215-239.

GONZÁLEZ DE MOLINA, Manuel (2001), 'Condicionamientos ambientales en las agriculturas españolas contemporáneas', in Josep PUJOL *et al.* (eds), *El pozo de todos los males,* Barcelona, Crítica, p. 43-93.

KOENIG, Nicholas (1994), *The Failure of Agrarian Capitalism. Agrarian Politics in the United Kingdom. Germany, the Netherlands and the USA, 1846-1919,* London/ New York, Routledge, 398 p.

MARTÍNEZ CARRIÓN, José Miguel (1986), 'Estatura, nutrición y nivel de vida en Murcia, 1860-1930', *Revista de Historia Económica,* IV, 1, p. 67-99.

MARTÍNEZ CARRIÓN, José Miguel (1988), 'Peasant Household Formation and the Organization of Rural Labour in the Valley of Segura during the Nineteenth Century', *Journal of Familly History,* XIII, 1, p. 91-109.

MILLÁN, Jesús (1991), 'Gran propiedad y pequeña explotación en el desarrollo del capitalismo agrario valenciano', in Ramón VILLARES (ed.), *Señores y campesinos en la Península Ibérica, siglos XVIII-XX,* Barcelona, Crítica, p. 7-33.

MILLÁN, Jesús & ZURITA, Rafael (1998), 'Elites terratenientes y tipos de caciques. La casa de Rafal/Vía Manuel entre la revolución liberal y la crisis del Antiguo Régimen', *Historia Agraria,* 16, p. 131-161.

PÉREZ PICAZO, Maria Teresa (1991), 'Riqueza territorial y cambio agrícola en al Murcia del siglo XIX. Aproximación al estudio de una contabilidad privada (1800-1902)', *Agricultura y Sociedad,* 61, p. 39-95.

PÉREZ PICAZO, Maria Teresa *et al.* (1992), 'Propiedad y explotación campesina en los regadíos del Segura durante el siglo XIX', in Ramón GARRABOU (ed.), *Propiedad y explotación campesina en la España contemporánea,* Madrid, MAPA, p. 163-164.

PÉREZ PICAZO, Maria Teresa (1997), 'Mercado internacional y estructuras agrarias en las agriculturas mediterráneas, siglos XIX y XX', in José MORILLAS *et al.* (eds), *Impactos exteriores sobre el mundo rural mediterráneo,* Madrid, MAPA, p. 389-433.

PÉREZ PICAZO, Maria Teresa (2000), 'Auge y decadencia del regadío tradicional en la región murciana, 1850-1960. Los logros de una agricultura de dominante campesina', in Carlos BARCIELA, Joaquín MELGAREJO (eds), *El agua en la Historia de España,* Alicante, Publicaciones de la Universidad de Alicante, p. 111-165.

PÉREZ PICAZO, Maria Teresa (2004), 'Economías campesinas y mercado. el caso de las huertas del Segura en los siglos XIX y XX', in E. LLEIDA (ed.), *Fires, mercat i mon rural,* Lleida, Diputació de Lleida, p. 15-42.

PÉREZ PICAZO, Maria Teresa (2005), 'El mercado de factores en la agricultura murciana durante el siglo XIX', *Investigaciones de Historia Económica,* 2, p. 39-75.

PÉREZ PICAZO, Maria Teresa, LEMEUNIER, Guy (2003), 'Los sistemas agrarios en la región murciana durante medio milenio (1500-2000)', in Constantino MARTINEZ GALLUR, Miguel Ángel ESTEVE SELMA, Mercedes LLORENS PASCUAL DEL RIQUELME (eds), *Los recursos naturales en al región de Murcia. Un análisis interdisciplonar,* Murcia, Universidad de Murcia, p. 170-213.

PÉREZ PICAZO, Maria-Teresa, PERCEVAL, Miguel Angel (1999), 'Ceux qui peinent et ceux qui prennent. Conjoncture et modes de faire-valoir dans les systèmes agraires du Sud-Est espagnol', *Histoire et Societés rurales*, 10, p. 47-81.

PONS, Anacleto (1993), 'Un huerto rodeado de secano. Informe sobre el cultivo del naranjo en el País Valenciano a fines del siglo XIX', *Noticiario de Historia Agraria,* 6, p. 37-59.

ROBLEDO, Ricardo, LÓPEZ, Santiago (eds) (2007), *¿Interés particular, bienestar público Grandes patrimonios y reformas agrarias ?*, Zaragoza, Prensas Universitarias Zaragoza, 581 p.

11. Specialisation in dairying. A rapid transformation in Denmark and a drawn-out process in Sweden

Carin MARTIIN

I. Introduction

Great changes took place in Danish and Swedish milk production in the mid-nineteenth century, generated by the well known contemporary expansion of international dairy markets, especially in England. Danish butter soon became world-famous and Swedish butter also found its way to the international arena. This transition was of the greatest importance to Danish and Swedish national progress, and for rural society in both countries. Behind the increasing volumes of dairy products were thousands of farms where traditional dairying had been changed in order to create a surplus for sale.

This article sheds light on the differences in pace and range of the specialisation process in Denmark and Sweden. In addition, it provides two Swedish examples at farm-level that give fragmentary glimpses from a more individual perspective. The time period stretches from the mid-nineteenth to mid-twentieth century, a long period that is explained by an early start in both countries but a drawn-out process in Sweden. The study of Denmark is based on existing historical studies whereas Sweden is studied through both historiographical and archival sources, many of which have been used in my earlier work (Israelsson-Martiin, 1998 and 2005).

The general history of the Danish dairy sector has been described by Claus Bjørn (Bjørn, 1982 and 1988), and the Danish ethnologist Bodil K. Hansen (Hansen, 1982 and 2006), amongst others (Henriksen, 1999, Svendsen & Svendsen, 2000, Henriksen & Hviid, 2002; Just 2003 and O'Rourke, 2005). An overview of the Scandinavian peasantry has been published by Orvar Löfgren (Löfgren, 1980) and one on the subject of cooperation, by Rydén (Rydén, 2004). An outline of Swedish dairy history

Agricultural specialisation and rural patterns of development, ed. by Annie ANTOINE, Turnhout, 2016 (*Rural History in Europe*, 12), p. 213-232.

BREPOLS PUBLISHERS DOI 10.1484/M.RURHE-EB.5.112269

is provided by Morell (Morell, 2001), while Swedish milk production and dairying have also been investigated (Sommestad, 1992 a and b; Niskanen 1995; Israelsson-Martiin, 2005).

While this article conforms the established perception of Denmark as an early and active participant in the specialisation of food production, it also shows how the Swedish dairy sector stands out as a mixture of an astonishingly low degree of specialisation and a correspondingly high degree of use in kind. These differences between the two neighbouring countries have been underestimated from a Swedish perspective. Certainly, Sweden experienced development in the milk sector, in the establishment of dairies, number of dairy cows and volumes of exported butter. But. in comparison with Denmark, the changes were modest and, most importantly, spread among the wider rural population much later and more gradually.

Focusing on Sweden, this interpretation stresses the importance of nuances to understanding cattle husbandry during the latter part of the Swedish agricultural revolution. It also highlights a need to extend the time period for the Swedish specialisation in dairying well into the 1930s. Moreover, it provides a better comprehension of an early twentieth-century Swedish rural society characterised by a low degree of specialisation in comparison with the Danish rural population, which was more involved in commercialization. Finally, these modifications may contribute to a better assessment of present-day power relations in what has now become frequent Danish-Swedish co-operation in the food sector. In this context, Denmark is generally regarded as the more purposeful and skilful party, while the Swedish collaborators keep a lower profile.

II. Commercialization and specialisation

Before making comparisons between the countries, specialisation in dairying will be briefly discussed more generally as they relate to the questions raised in the introduction of this book. The words 'commercialization' and 'specialisation' when applied to dairy production can often be used for the same process, a widespread expansion of milk production intended for sale. But although they are closely intertwined, the two words do differ in some aspects. It is possible to be commercial without being specialized; varying kinds of commodities can be produced and sold from the same unit, specialisation involves more active choices, choices to concentrate resources in order to produce one or a few goods and sell them, while other kinds of production are given up and replaced by purchases.

In Denmark and Sweden, commercialization of milk required increasing the quantity produced to volumes exceeding subsistence levels. In order to create such a surplus, resources like arable land and labour had to be concentrated more on dairy production than before. As most Danish and Swedish farms were small, increased feed production on arable land brought about lower production of commercial crops. Especially in the Swedish context, feed production on arable land included not only oats and barley but also hay and sugar beets. According to previous studies increased amounts of hay were of great importance for the initial efforts to increase milk volumes. Previously, cattle used only for subsistence had been severely and widely malnourished, being fed basically on straw and other by-products of the arable, meadows and wooded land.

As undernourishment was a key factor in low milk yields, modest changes such as replacing straw by hay had already brought about radical improvement and the possibility of creating a surplus for commercial purposes (Israelsson-Martiin, 2005: 276). This means that the early stages of dairy specialisation in Sweden should not be interpreted as a simple reflection of grain price levels. Thus, the Swedish point of departure was somewhat different in regard to the feed situation than that in the Low Countries, where access to cheap grain was a key factor for specialisation in dairying (Brenner, 2001: 230).

Another question is whether specialisation in milk meant giving up producing of necessities for use in kind. This subject reveals interesting differences between Danish and Swedish strategies. As Figure 11.4 (see below) shows, the Danish degree of specialisation was much higher than the Swedish, which may be described as 'half-hearted' well into the mid-twentieth century. The Swedish tendency to combine commercial and subsistence production refers to fresh milk, but an even more striking example is the continuing but not very rational economic choice to grow potatoes, widely practised to this day. Conversely, other customary products were successively given up in Sweden, such as traditional handmade textiles, a well known tendency discussed by de Vries (de Vries, 1994: 256).

The introductory text asks if specialisation leads to mono-activity. Although both Danish and Swedish farming continued to have the character of mixed farming, extended milk production resulted in fewer activities and far-reaching structural changes at farm level. Increased dairy production helped put more emphasis on farming, part-time farmers became full-time farmers, while a number of complementary pursuits were more or less abandoned. The process of commercialization and specialisation also included a re-definition of milk production: formerly a female, household-related task, it became an important part of farming, receiving increased male attention (Sommestad, 1992a and 1992b: passim; Israelsson-Martiin, 2005: 249-50).

A final more general question deals with the need for capital in order to bring about specialisation. In the case of milk, the answer is that dairy production could be, but was not always, specialised without investments and going into debt. In principle milk production could be expanded within an advanced organic farm economy, as theoretically described by Edward Wrigley (Wrigley, 1988). Calves could be bred from dairy cows, and grass and clover could be reproduced with almost no purchases. A similar observation has been made by Michael Winstanley in his study of small-scale dairy producers in Lancashire (Winstanley, 1996: 178). Furthermore, frequent and stable payment from dairy plants made it possible to buy input resources like concentrates and seed almost in parallel with improvement in production. In addition the cowshed might be slightly expanded without much investment and machines for more efficient harvesting of feed stuff could be borrowed temporarily (Israelsson-Martiin, 1998). Such strategies, where cash income from milk was one step in front of expenditures, made it possible to develop subsistence dairy farming into small-scale specialisation. Swedish figures showing a relatively high proportion of dairy herds with fewer than four milch cows indicate that such strategies were carried out in practice (Swedish Dairy Statistics, 1913-39). But although some degree of specialisation could be attained with very little money, there is no doubt that the processes of commercialization and specialisation in dairying contributed to the wide expansion of cash economies, first and foremost in Denmark but also in Sweden.

III. Rural society in Denmark and Sweden

Well into the last decades of the nineteenth century, Denmark and Sweden had a large rural population with an urban population of less than 25 and 10 per cent respectively (Bjørn, 1988: 12; Swedish Statistics, BiSOS A). By this time, industrialisation and urbanisation had begun, and had contributed to an expanding domestic commercial demand for food. According to this timetable, the domestic Danish and Swedish food markets obviously did not expand until decades after butter exports rose after mid-century.

The Danish countryside was relatively densely populated, with an open, arable landscape, peasant villages, some estates and a number of small towns. This structure provided reasonably short distances between people, and between dairy producers and potential customers. Moreover Denmark, situated on a peninsula and islands, had access to good seaways, while communications in the vast Swedish inland were hampered by few and bad roads.

Compared with Denmark, the Swedish countryside can be described as dominated by forestland and sparsely populated. There were however great differences between the various parts of Sweden: the south was reminiscent of Denmark, there were forest-dominated regions alternating with more fertile arable areas in the central regions, and cold and barren conditions in the far north.

Although its natural and cultural geography were favourable, history had held back Danish agricultural and rural development for centuries. Although Swedish yeomen had enjoyed a considerable freedom, their Danish equivalents lived under more severe pressure from the nobility until the late eighteenth century. Comparisons between the rural and agricultural changes in Denmark and Sweden in the early nineteenth century indicate that the new freedom of the Danish yeoman provided an important springboard to speedy and effective rural development. One example is the rapid reconstruction of the Danish arable landscape, completed within decades, while the corresponding Swedish process lasted for more than one hundred and fifty years (Bjørn, 1988: 14; Gadd, 2000: 273 and 300). Danish capability is well known: van Zanden for example argues that Denmark in 1870 was one of four European countries on 'the efficiency frontier', while Sweden belonged to the 'medium level… in a circle around the core' (van Zanden, 1991: 219).

Although the Danish and Swedish rural populations had different historical backgrounds, the socio-economic structure and size of farms and holdings were relatively similar during the period under study. Danish holdings had more arable than Swedish holdings, but both countries can be described as possessing small- scale agriculture in parallel with some larger farms and estates. By the time of the First World War, 55 per cent of Danish holdings had fewer than ten hectares of arable land, 44 per cent had between 10 and 60 hectares, and 3 per cent had 60 hectares or more. In Sweden as much as 90 per cent of the holdings had fewer than 10 hectares of arable, while only 7 per cent had between 10 and 60 hectares, and 2 per cent had 60 hectares or more (Danish Historical Statistics, Swedish Statistics N, 1913, respectively).

The small-scale character of the Danish and, a fortiori of the Swedish holdings, was even more pronounced in regard to dairy cattle husbandry. The number of cows per herd was of great importance in creating a marketable surplus. According to my earlier studies, one or two cows per herd were common in Sweden; it must have been hard to create any surplus at all from that level (Israelsson-Martiin, 2005: 263). A lack of Swedish statistics makes it difficult to compare Danish and Swedish herd sizes, but I would estimate that Sweden had twice as many herds as Denmark, and that the Swedish herds were about half the size of the Danish (Bjørn 1982: 543; Israelsson-Martiin, 2005: 95).

IV. Traditional dairying in Denmark and Sweden

Notwithstanding the demographic, geographic and historical differences, dairy cattle were traditionally spread all across the Danish and Swedish countryside; dairy cows were found on all large farms and small holdings. The animals were traditionally kept for several purposes such as for milk, meat, calves, draught power, status and so forth, and their milk was chiefly intended for the internal needs of the holding. When and if the need for human and animal milk consumption was met, the surplus was converted into various dairy products.

Traditionally, dairying on Danish estates and large farms has been described as a regular farmhouse production of cheese and butter, routinely sold at local and regional markets or to city-dwellers in Copenhagen (Højrup 1972: 75; Pedersen 1944: 45). On smaller Danish farms and holdings, the ordinary kitchen was the usual place for a more limited preparation of the finer dairy products, which was the responsibility of the farmer's wife, daughters or maids (Højrup, 1972: 70-75; Feilberg, 1898: 118; Reimer, 1912: 208). Part of the surplus from such small-scale production could be sold, but some cheese and butter were kept for important social purposes, such as gifts at weddings and funerals (Pedersen, 1944: 42). Milk or milk products were also given to the rural poor as part of the social responsibilities of local society. Moreover, milk and milk products could be used as loans or as exchange commodities when one household had a surplus and the other a deficit of milk (Pedersen, 1944: 38; Reimer 1912: 214).

These characteristics were applicable to rural societies in both countries, and had formed a natural part of rural Danish and Swedish societies for centuries (ULMA, questionnaire n° 1). The relative proportion of these elements is difficult to estimate but several features point towards larger and more frequent surpluses for sale in Denmark than in Sweden. Among the indicators of higher surpluses in Denmark were: its geography, its more favourable climate, its larger herd sizes and access to better feed stuff and pastures. As we have seen, most Swedish dairy cattle were badly undernourished in winter time and kept on woodland during parts of the year spent outdoors; forest land was seldom found in Denmark.

Despite many similarities in the broad outlines of traditional Danish and Swedish dairying, there were some differences. Among other things, Swedish narrative sources indicate that Swedish rural women, more often than the Danish, regarded dairy products as something to be used in kind, not to be sold (NM 60 and ULMA n° 1). According to these indications, the ambition to turn dairy products into money may in general have been less in Sweden. Additionally, both countries had remote areas, where traditional dairying skills probably remained long after the general commercialization.

From the 1830s onwards, Danish estate dairy production received moreattention, and skilful dairymaids from Holstein were employed in order to improve the quality of dairy products and to teach good dairying. According to the Danish ethnologist Bodil K. Hansen, the primary motive for increased Danish dairying was related to fear that soils would be impoverished as a result of ongoing intensive grain production (Hansen, 2006: 178; Hansen, 1982: 227). The initiative proved to be a fruitful start to what was to become a very successful specialisation in butter. Even though Denmark benefited most from the import of skills, the principles of 'the Holstein dairy' left their mark in Swedish dairying as well, often through the employment of Danish dairymaids in Swedish dairies (Ingers, 1956: 78; Regional five year reports, BiSOS H; Summary 1881-85; Säve, 1914-1917: 238).

V. Specialisation in dairying at national levels

By the 1860s and 1870s a conscious strategy for developing dairy production was actively carried out by individuals and organisations, and at the political level in Denmark. Animal production in general and milk in particular, were designated as the future for Danish agriculture (Bjørn, 1982 and 1988). The period from 1860 onwards is described as the initial phase of a systemic shift from commercial grain to animal production, and as a period when commercial dairying became relatively well established among larger farms. The focus on dairying escalated in the 1870s but the real breakthrough came during the 1880s and 1890s. By this time, the well-known Danish co-operative dairy movement had rapidly and successfully been built up, and before the turn of the century almost all the Danish holdings seem to have been connected to co-operatives. From then on, Danish dairy production seems to have been more or less fully commercialised and Danish specialisation in butter production soon became a model for other countries (Gráda, 1977: 298).

As for Sweden, a regional author of the time sheds light on Swedish admiration for 'Denmark, whose dairy-practicing population always has been a model for the Swedish dairy system' ('Danmark, hvars mejeriidkande befolkning städse varit en föresyn för det svenska mejeriväsendet') (Nyberg, 1914: 245). Swedish agricultural representatives were well aware of Danish inspiration and developments, and the results of experimentation were disseminated by such institutions as the Royal Swedish Academy of Agriculture in Stockholm. We can observe some of the early development of the Swedish dairy sector through a regional example in the papers of a dairy advisor officer, John Nyberg from the Swedish county of Jönköping. It is clear from his writings that Nyberg was close to agricultural life as it was practised in the real world, unlike many other authors of the time. In addition, Nyberg appears in the diary that is the main source in the first farm example, 'Little Mölarp', which he visited in November 1907 in order to discuss the local milk control association (Israelsson-Martiin, 1998: 25). John Nyberg mentions that the influences from Denmark and Holstein had reached the county of Jönköping as early as the 1840s. He also describes a small regional dairying school in 1859 and the employment of a 'county dairy-maid' during the 1860s.

Notwithstanding these attempts, Nyberg claims that the proper start of dairying came a little later, along with what was called 'the ice-method', that is, the Schwartz method. This resulted in several new private dairies in the county, one of which sold dairy products directly to England, Copenhagen and Gothenburg (Nyberg, 1914: 238-243).

Despite a relatively good start, only one or two decades behind the Danes, the Swedes seem to have been slower to react and sometimes reverted to earlier habits. The latter happened for example when the rapid spread of the small-scale separator made it possible to avoid time-wasting transport of farm products to factory dairies. Swedish national dairy statistics do not cover the 1880s and early 1890s, but figures from the Agricultural Society of Jönköping reveal closed dairies in the region when the use of the small separator became common (Nyberg, 1914: 244). Such a 'reversion', does not seem to have occurred in Denmark, as far as we can see from the historiography. The pattern of trials followed by failures and periods of recovery also reappeared during the First World War, when direct sale became extremely attractive, and around 1930, when the price of milk fell dramatically. By contrast, the Danish transition was more or less linear and already complete by the end of the nineteenth century, just as the Swedish process was beginning. By 1890, only about 10 per cent of the Swedish herds were connected to a dairy and in 1913/15 about 25 per cent. In the 1930s half of Swedish herds still produced primarily for use in kind, but after the introduction of 'the dairy regulation' in 1932, interest in concentrated production rose, and in 1938 about two thirds of all dairy cows were linked to a dairy (Swedish Dairy Statistics, 1913-1939, comments, p. 4).

V.1. An overall presence of dairy cattle in both countries

Despite the fact that Denmark was already on the way to becoming a leading dairy producer with Sweden a step behind, dairy cattle were overwhelmingly present in both countries. This feature grew even stronger during the latter half of the nineteenth century and the beginning of the twentieth. During this period, the number of dairy cows increased from 0.8 to 1.2 million in Denmark and from 1.2 to 1.9 million in Sweden.

Comparisons between the human population and national stock of dairy cows in each country show similar and almost linear increases of people and dairy cattle in both. During the period studied, the proportion between the number of people and dairy cows was stable: slightly more than two inhabitants per cow in Denmark, and a little more than three inhabitants per cow in Sweden (Danish Historical Statistics; Bjørn, 1982: 541; Swedish official statistics). By contrast, England & Wales had 12 persons per dairy cow in 1867 and 22 in 1939 (Mitchell, 1962). A combination of low urbanization rates in both Denmark and Sweden (25 and 10 per cent respectively) and widespread milk production among the many rural inhabitants, reduced the incentive to specialize in commercial dairy production for domestic markets. Who wanted to pay cash for milk and milk products when almost everybody had access to milk in kind?

Figure 11.1. Number of dairy cows in Denmark and Sweden, 1861-1939

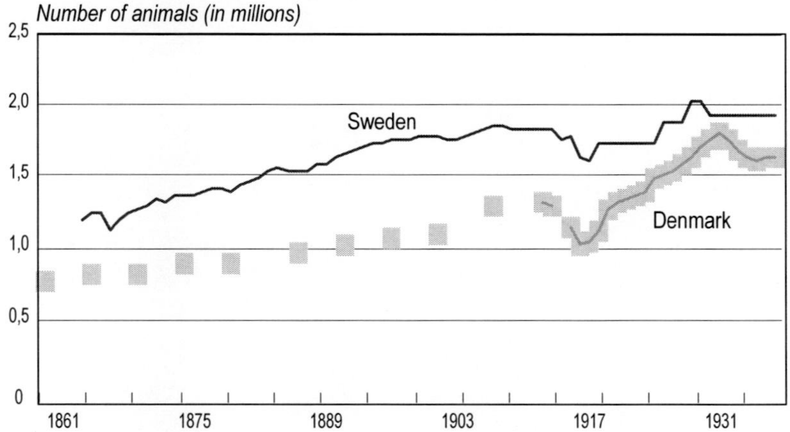

* The Danish figures were given in five-year intervals until 1914.

Sources. Bjørn (1982: 541; Denmark, 1861-1939) Swedish statistics, BiSOS and SOS respectively, Arable farming and husbandry; series N (Sweden, 1865-1939).

V.2. Similar total volumes but different priorities

Although few Danes and few Swedes were interested in buying milk, English consumers emerged as potential customers and soon became the target for early specialisation in dairy products. Later on in Sweden, by the turn of the century, urban demand expanded and attracted more attention from producers than the export market. By contrast, the domestic Danish market remained of secondary importance for Danish butter production. This difference in export ambitions helps understand the increasingly divergent patterns seen in Figure 11.2. Danish butter exports were almost eleven times higher by the late 1930s compared with the late 1870s while Swedish butter exports increased, but only six-fold.

Despite the widening differences between exported quantities of butter, overall Danish and Swedish milk production was relatively equal. The total Danish milk volume increased from 2.4 billion kilos in 1900 to 5.3 billion in 1939 (Bjørn, 1988: 545). We do not have statistics for total Swedish milk production that also take account of production for use in kind. I would estimate however that the total Swedish production rose from about 1.5 billion kilos per year in the 1860s to about 4.6 billion kilos in 1936 (the latter figure from Mjölkpropagandan 2 – 1937: 42).

Figure 11.2. Danish and Swedish butter exports, 1874-1939

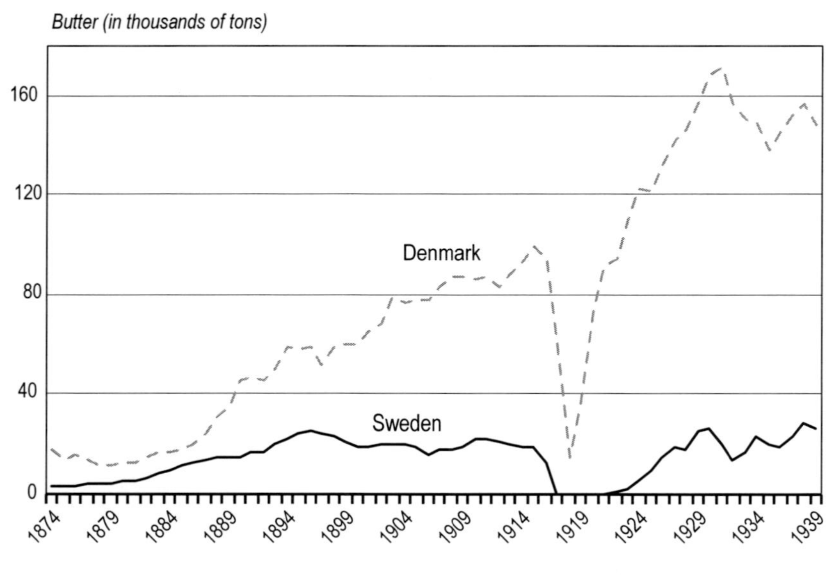

* The Danish figures include transit trade until 1897 (including transit from Sweden). Six year-averages, 1874-1879 to 1934-1939.

Sources. Bjørn (1982: 549) resp. Swedish statistics, BiSOS and SOS, Foreign Trade, series F.

Figure 11.3 shows similar volumes of total production in Denmark and Sweden. The figures are based on milk delivered to dairy plants plus estimates for volume of use in kind. This means that a large proportions of the Swedish production is an estimated figure (Comments to Official Dairy Statistics 1933-1938).

How were such large Danish exports possible? Firstly, Denmark had much tighter links to dairy plants. Secondly, the dairy plants specialised in butter production.

Thirdly, and quite contrary to Sweden, Danish domestic consumption of whole milk was minimized in order to maximize butter exports.

A deeper understanding of the reasons for these differences between Denmark and Sweden would go beyond the aims and possibilities of this article. But it would not be unfair to compare the Danish approach with that of an anthill, whose inhabitants regarded a maximum butter export as a sort of national duty. Earlier in this article we considered present specialisation as an active choice to give up something in order to gain by focusing on fewer products. In this perspective the commercialization of Danish milk demonstrates extreme specialisation, as demonstrated by widespread consumption of margarine on Danish farms. By contrast, the Swedish approach may be described as less activist and the interest in trade as more reluctant, mirrored in frequent consumption of milk and butter at Swedish holdings. Figure 4 illustrates the principal differences, and the much heavier emphasis on specialisation in Denmark.

Figure 11.3. Estimated total milk production in Denmark and Sweden, 1871-1936

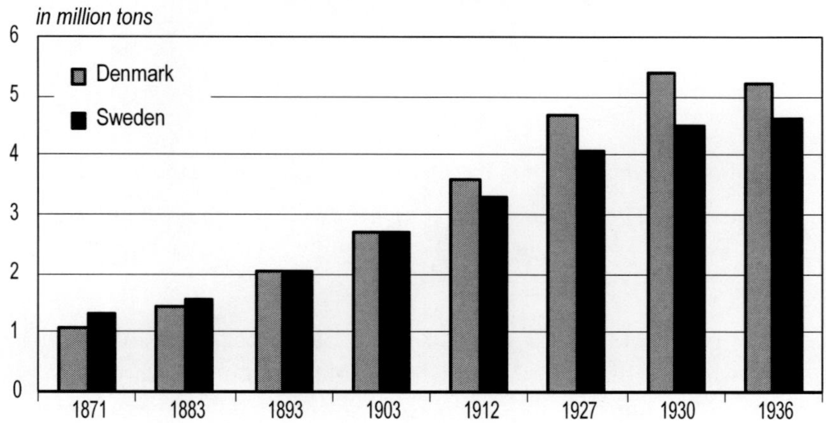

Sources. Ellebrecht (1931: 17; Denmark) and the journal *Mjölkpropagandan*, 3, 1931: 40, (article by Höijer, Statistics Sweden) and *Mjölkpropagandan,* 2, 1937: 42f (Sweden).

As Figure 11.3 shows, considerably larger shares of the total Danish milk volumes reached dairy plants and higher proportions of products from these plants reached the export market. In Sweden, more milk was kept in rural households, both in households that were also producing for dairy plants and those without links to a dairy. Moreover, the figure for Sweden indicates the small volume of exports, and the large volume that was ‘recycled’ in rural society. We lack figures in this area, but Table 11.1 presents the approximate destination of all whole milk produced in Denmark and Sweden in

1928. Table 11.2 shows the distribution between domestic use and export. The figures are based on volumes that reached dairy plants.

Table 11.1. Destination of milk production by volume, Denmark and Sweden c. 1928

	Milk used in kind %	*Milk to dairy plants* %	*Total* %
Denmark	5	95	100
Sweden	40	60	100

Sources. ELLBRECHT (1931: 19; Denmark) resp. *Mjölkpropagandan,* 9, 1929: 171 (article by HÖIJER, Statistics Sweden).

Table 11. 2. Destination of milk treated in dairy plants by volume, Denmark and Sweden c. 1928

	Domestic use %	*Export* %	*Total* %
Denmark	12	88	100
Sweden	89	11	100

* The export of cheese was marginal by this time, and has been left out of account.

Sources. ELLBRECHT (1931: 19; Denmark) resp. *Mjölkpropagandan,* 1, 1929: 26 and my accounts, based on the Swedish butter export according to (Danish Historical Statistics resp. Swedish Statistics, Foreign Trade, series F).

Table 11.1 and Table 11.2 would seem to show that domestic Danish milk consumption was very limited. A closer investigation into Danish consumption of such products as skimmed milk and margarine may help further understand the Danish strategy of extreme specialisation in butter.

VI. A Danish farmer sells milk

As mentioned earlier, geographical circumstances explain some of the differences in patterns of production and delivery to dairy plants. But natural environment cannot on its own account for all of the huge differences between Denmark and Sweden appearing in Table II.1. and further discussion of the reasons is needed. In this article different attitudes to commercialization and specialisation are suggested as an influential factor. This understanding developed in parallel with a careful reading of ethnological questionnaires; the farmers' journal 'The Milk Propaganda'

(*Mjölkpropagandan*) and archival material regarding the second farm-example in this article. All indicated that Swedes were less interested in selling maximum quantities of dairy products. Further studies of Danish literature have strengthened the opposite impression, describing Danish peasants and farmers as enthusiastic about commercial dairying and improving their volume and quality.

Among the driving forces that contributed to these attitudes in Denmark, some some reflections about identity and milk need to be made to start with; we can then go on to look at aspects of organization. A general picture of the Danish farm during the period studied shows a unit involved in a modern project, where the farmer is an actor in a successful international business, in which dairy products played a key role. Although Danes also kept many of their rural traditions, the links to the past are more evident in the Swedish context. The Swedish farmer is not portrayed as primarily relating farming to the economy or to milk as a commercial commodity. Rather, the Swede appears to be a trustee, proud to manage ancestral soil without excessive involvement with the outside world.

When it comes to organization, it must be emphasised that milk is perishable and has to be processed both rapidly and carefully. Consequently, many advantages could be gained by collaborative organizing to move output from the many far-flung holdings to operate a dairy plant for processing and to then move the product to costumers. There were efforts to organize the management of milk in both countries, but in Sweden the result was unimpressive and characterized by small-scale experiments and trial and error rather than by purposeful projects. The well-known Danish dairy cooperative network was established rapidly, while the corresponding buildup in Sweden, which later also had very strong dairy cooperatives, did not reach a similar level until after the Second World War.

VII. Specialisation in dairying at the farm level

The discussion on specialisation now turns to farm level and two examples from Sweden. This study is based on non-print sources, and describes two different social contexts during the late nineteenth and early twentieth centuries. The first example comes from a middle-sized farm, where regular deliveries of fresh milk started in 1895; the second case-study covers the development of specialised butter production on a middle-sized Swedish estate. The first case, Little Mölarp, was thoroughly investigated in an earlier study, though with another focus (Israelsson-Martiin, 1998). The second example comes from the Krusenberg estate and draws on material from the Regional Archives in Uppsala, Krusenberg archive and (Israelsson-Martiin, 2005), an important source used for my doctoral dissertation.

On both the farm and the estate, milk production was commercialised early on, earlier than in the general pattern traced in this article. In the respective time periods, the two examples show similarities with the Danish situation; they were both commercial by the beginning of the twentieth century. One difference, however, was that the Swedish farm and estate were both dependent on private dairy plants, while their Danish colleagues were already connected to a strong, nationwide co-operative movement.

VII.1. Little Mölarp, daily deliveries from one day to the next

The history of Little Mölarp appears in roughly fifty years of daily notes in a diary written by the yeoman Frithiof Johansson between 1875 and 1923. The notations cover a wide range of subjects, including fragments about the commercialization of the farm economy.

Little Mölarp was a typical medium-sized farm, situated in the northern part of the county of Jönköping, halfway between Stockholm and the southern Swedish border. Traditionally, animal production was very important in the area, with a high density of animals per hectare of arable. Before fresh milk began to be delivered regularly in 1895, the cash income from the animals was chiefly derived from the sale of oxen, the principal interest of the farmer.

Notations from earlier decades show an occasional sale of cream, but this seems to have been a sporadic activity, occuring infrequently and on a very small scale. The diary-writer's cream deliveries shed light on the relatively common occurrence of so-called cream-dairies in Sweden. According to a comment in a regional five-year report, cream dairies were a way to solve 'the dairy question in forest areas' (Swedish statistics, series H). Of all deliveries to the Swedish dairy plants in 1913, 3 of 1,200 million litres, or 0.25 per cent, were cream (Swedish Dairy Statistics 1913, p. 13). [But this was not the situation earlier in the century]. Producers for cream dairies skimmed the milk at home, used the skimmed milk as food and animal feed, and delivered the more concentrated cream as a 'semi-finished product' to the dairy. The cream was less vulnerable than the fresh milk and could be delivered every eighth or fourteenth day (Nyberg, 1914: 244). Hence, selling cream instead of milk saved time, work and draught power for the producers. Moreover, it broadened the area from which the cream could be bought and offered commercial possibilities for those who lived too far from dairy plants to be able to manage a daily delivery. In addition, the remaining skimmed milk could generate a small scale specialisation in calves or pigs in some areas. This might be a strategy for farms connected to butter producing plants, while those delivering for cheese production had access to lesser amounts of skimmed milk.

The change in deliveries of fresh milk from Little Mölarp indicates a switch-over at the dairy, from cream to fresh milk. This in turn indicates that the need for butter production could be filled from a smaller area, small enough for fresh milk to be transported daily. This should however not be understood as the result of an more cattle herds, because almost all rural inhabitants already had cows. The explanation for an increased supply of fresh milk, as occurred in the small dairy plant near where the author of the diary lived, is to be found rather in increased efforts to create a marketable surplus. Such a surplus may have been created through one or more combinations: more milk per cow; bigger herd sizes and more awareness of the uses of milk in kind (further details on means of production and the socio-economic situation in Swedish cattle husbandry at this time are included in Israelsson-Martiin, 2005).

For the diary writer, the start of fresh milk deliveries in 1895 was the big step towards commercialization. From one day to the next, milk would now always be delivered to the dairy. Likewise, from one day to the next, cash money would now regularly be paid out, throughout the year. From then on the household became less dependent on the price of oxen and harvests of grain. Compared with earlier years, the five brief notations in 1895 and eight in 1896 made frequent references to milk:

1895:

14 March: for milk (Oberga), 7.61 Swedish crowns

21 June: Went to Tranås to get money for milk, 32.82 crowns 12 July: Was paid for milk from the dairy, 29.16 crowns

8 November: milk, 27.39 crowns

22 December: the diary 22.06 crowns

1896:

14 February: for milk from Oberga, 29.70 crowns 18 June: From the dairy, 71.40 crowns

16 July: For June, from the dairy 56.49 crowns 30 July: 60.65 crowns for milk from Oberga

11 September: The dairy, 80 litres as 78, 62.48 crowns

9 October: The dairy 78.794 litres, 61.93 crowns 24 November: October 30.60 crowns, 382½ litres 11 November: from the dairy, 298, 23.84 crowns

The small dairy plant was situated at the relatively large 'Oberga' farm, several kilometres from Little Mölarp. Frithiof Johansson mentions both Oberga and the small town Tranås, situated in the other direction from Little Mölarp. It seems that milk was delivered westwards to Oberga but the payment was made in Tranås, probably at the bank, which the author frequently visited for varying reasons.

In January 1897, the notations about payment from the dairy suddenly cease. A possible reason is that the income from the milk had become so regular and important that the author decided to keep another book for it, a book that no longer remains.

Although the sporadic notes about quantities and payment only refer to two years, other notations mirror a continued sale of fresh milk. The daily milk transport is frequently mentioned in the later years, when a son was running the farm. Frithiof Johansson now took on a new role as an assistant who often made the daily transport of fresh milk to the dairy. Despite the above comment on difficulties caused by long and frequent shipments, I would suggest that the relatively short trips from Little Mölarp to the dairy had become a kind of a pleasure for the now aging, still rather sociable yeoman.

It is interesting to register the effects of the regular income from the fresh milk, which immediately set off investments in the cowshed. As early as in June 1895, a new feeding table was built there. During the summer and the following years, bran and oil-cakes were purchased, water installed in the cowshed and so forth. Moreover the family's private consumption increased with, among other things more railway journeys, visits to the dentist and purchases of clothes and books.

The history of the diary shows how the transition to a specialised production of fresh milk for sale came about almost from one day to the next. The quick change was, however, preceded by occasional deliveries of small surpluses. The sources do not provide any information on milk volumes, but the number of cows was surprisingly stable, about 10 cows, between 1867 and 1941 (Primary statistics, 1867; notations in the diary 1877; estate inventory, 1901 and Primary statistics 1941). Diary notations about the purchase of oil-cakes, bran and effective harvest technology, as well as membership of 'the milk control' do however point to higher milk yields per cow and consequently rising surpluses for sale.

VII.2. Krusenberg, decades of building up

While the commercialization and specialisation at Little Mölarp resulted in a rapid change in 1895, the Krusenberg estate displays a slower process. Krusenberg is situated about 60 kilometres north of Stockholm and 15 kilometres from Uppsala, with waterway and railway connections to Stockholm. Despite the relatively short distances to urban areas and consumers of fresh milk, dairy production was concentrated on butter.

Archival documents show a shift from grain to milk and butter starting in the 1860s. During this decade, the relative income from the cowshed (milk, meat and the like) rose from 7 per cent of total agricultural income in 1860 to 20 per cent in 1869. Moreover, investments were made in a dairy and in a large new cowshed during the 1870s. The stable and dairy journals give evidence of a successive increase in dairy production. Despite access to the estate dairy and to the new cowshed, the extension was a drawn-out process that lasted until the 1890s. By that time, the number of dairy cows had increased from about 30 to more than 80 and milk production had increased five-fold. Dairy cows now dominated farming at the estate. In 1896 as much as 80 per cent of the arable (excluding fallow) was used for animal feed, most of it to meet the needs of the dairy cattle.

Apart from the slow pace of the commercial expansion which has just been described, we also find a remarkable combination of commercial butter production with a much use in kind on the estate. Surprisingly large volumes of whole and skimmed milk were used to pay farm labour all through the period under study. Thus the specialisation in butter production did not mean that all the milk from this particular farm went to commercial sale. On the contrary, commercialization was balanced towards a conscious strategy of continuing to use of milk in kind (Martiin, forthcoming).

VIII. Concluding discussion

The article presents specialisation as an active choice to give something up in order to gain by focusing on fewer products. The process of specialisation in dairying is often driven by commercialization and the words specialisation and commercialization may be closely intertwined. Yet the two words differ in some ways: commercialization is possible without specialisation while the opposite is not realistic in practice. Moreover, specialisation involves more active choices; choices to concentrate resources in order to produce one or a few goods and sell them, while giving up other kinds of production and replacing them by purchases.

The article presents the Swedish process of specialisation as much more hesitant than the Danish, which appears as rapid, widespread, straightforward and completed within about two decades. Some explanation of the difference is found in varying natural circumstances like geographical distances and population density, but, it is also suggested, varying Danish and Swedish attitudes to the sale of milk also affected the degree of specialisation: the creation and marketing of a dairy surplus required new ways of thinking and new ways of working.

Although specialisation in dairying appeared earlier in Denmark, Swedish discussions and initiatives seem to have been relatively up-do-date by the middle of the nineteenth century. It soon became evident however that Sweden generally tendened to stay one step behind their small but purposeful neighbour to the south. While the Danish process escalated, the Swedish process slowed down. Danish producers were almost totally connected to dairy plants before the turn of the century, but the corresponding Swedish process was still not complete by the outbreak of the Second World War. That was nearly fifty years after the total breakthrough of dairy specialisation in Denmark. In Sweden, less interest in deliveries to dairy plants and a generally more hesitant approach contributed to these differences. Moreover, the Swedish milk sector continued to be characterized by a combination of use in kind and deliveries to dairy plants, a sign of incomplete commercialization and a low degree of specialisation.

Archive material

The Regional Archives in Uppsala: Krusenberg archive ; Primary statistics, (farms and parishes in Ärlinghundra hundred, Sweden).

The Regional Archives in Vadstena : Primary statistics (farms and parishes in Norra Vedbo hundred, Sweden).

The Swedish Institute for Language and Folklore: Questionnaire ULMA, No. 1, Dairying.

The Swedish Nordic Museum: Questionnaire NM 60, Cattle husbandry.

Frithof Johansson's diary 1875-1923. (A reprint is kept by the author).

Printed sources

Danish Historical Statistics 1814-1980. See Christensen & Johansen, 1985.

The Milk Propaganda (Mjölkpropagandan); journal, 1924-1939.

Swedish Dairy Statistics, 1913-1939.

Swedish Statistics, BiSOS and SOS, Population, series A.

Swedish Statistics, BiSOS and SOS, Foreign Trade, series F.

Swedish Statistics (Regional five-year reports), BiSOS, series H.

Swedish Statistics, BiSOS and SOS, Arable farming and husbandry, series N.

Bibliography

BJØRN, Claus (ed.) (1988), *Det danske landbrugs historie III, 1810-1914*, Odense, Landbohistorisk Selskab.

BJØRN, Claus (1982), *Dansk mejeribrug 1882-1900,* Odense, Landbohistorisk selska.

BRENNER, Robert P. (2001), 'The Low Countries in the Transition to Capitalism', *Journal of Agrarian Change*, vol. 1, 2, p. 169-241.

DE VRIES, Jan (1994), 'The Industrial Revolution and the Industrious Revolution', *The Journal of Economic History*, 54, 2, p. 249-270.

ELLBRECHT, G. (1931), *Dansk mejeristat,* Published by 'Vort samfund', Köpenhamn.

FEILBERG, H. F. (1898), *Dansk Bondeliv, saaledes som det i Mands Minde førles. navnlig i Vestinlland,* Kjøbenhavn, G. E. C. Gad, X-394 p.

GADD, Carl-Johan (2000), *Den agrara revolutionen 1700-1870,* in *Det svenska jordbrukets historia* (*History of Swedish Agriculture*), 3, Stockholm, Natur och kultur.

HANSEN, Bodil K. (2006), *Familie- og arbejdsliv på landet ca. 1870-1900,* Auning (Denmark), Landbohistorisk Selskab.

HANSEN, Bodil K. (1982), 'Rural Women in Late Nineteenth-Century Denmark', *The Journal of Peasant Studies,* 9, 2, p. 225-240.

HENRIKSEN, Ingrid (1999), 'Avoiding Locking-in: Cooperative Creameries in Denmark 1882-1903', *European Review of Economic History,* 3, p. 57-78.

HENRIKSEN, Ingrid, HVIID, Morten (2002), 'Legal Institutions and Performance. Monitoring Agreements in the Early Danish Sector', Paper presented at the EAAP conference, 2001, (the European Association for Animal Production).

HØJRUP, Ole (1972), *Landbokvinden. Skive. Højrup, Ole (1975), Landbokvinden : rok og kærne, grovbrød og vadmel*, København, Nationalmuseet, 277 p.

INGERS, Enoch (1956), *Bonden i svensk historia*, 3, Stockholm, Lantbruksförbundets tidskriftsaktiebolag.

ISRAELSSON-MARTIIN, Carin (2005), *Kor och människor,* Published dissertation, Swedish University of Agricultural Sciences, Uppsala.

ISRAELSSON-MARTIIN, Carin (1998), *Frithiof Johanssons dagböcker 1875-1923. Fem decennier av agrar och urban förändring. Essay in Agricultural and Rural history,* Uppsala, Swedish University of Agricultural Sciences.

JUST, Flemming (2009), 'From crisis to cream: The Scandinavian food system in the interwar period', in Yves SEGERS, Jan BIELEMAN & Erik BUYST (eds), *Exploring the food chain: Food production and food processing in western Europe, 1850-1990,* Turnhout, Brepols Publishers, CORN publication series, 11, p. 213-228

LÖFGREN, Orvar (1980), 'Historical Perspectives on Scandinavian Peasantries', *Annual Review of Anthropology,* 9, p. 187-215.

MITCHELL, B. R. (1962), *Abstract of British Historical Statistics,* Cambridge, Cambridge University Press, I.

MORELL, Mats (2001), *Jordbruket i industrisamhället 1870-1945,* in *Det svenska jordbrukets historia,* (History of Swedish Agriculture), 4, Stockholm, Natur och kultur.

NISKANEN, Kirsti (1995), *Godsägare, småbrukare och jordbrukets modernisering. Södermanlands län 1875-1935,* Published dissertation, University of Stockholm.

NYBERG, John (1914), 'Mejerihandteringen inom Jönköpings län 1814-1913', in Gunnar SÄVE (ed.), *Jönköpings läns Kgl. Hushållningssällskaps historia 1814-1913,* 1, Jönköping, Sweden, p. 237-255.

Ó'GRÁDA, Cormac (1977), 'The Beginnings of the Irish Creamery System, 1880-1914', *The Economic History Review,* 30, 2, p. 284-305.

O'ROURKE, Kevin H. (2005), 'Late 19th Century Denmark in an Irish Mirror: Land Tenure, Homogenity and the Roots of Danish Success', Paper from the Department of Economics and IIIS (School of Social Sciences and Philosophy) at Trinity College Dublin, 56 p. Online : http://www.tcd.ie/Economics/staff/orourkek/offprints/dartmouthfinaljan2005.pdf

PEDERSEN, Anna (1944), *Fra bondestue og stegers. Obtegnelser fra Stevns og Bjeverskovs herreder',* Copenhagen, Ejnar Munksgaard, 151 p.

REIMER, Christine (1911), *Nordfynsk Bondeliv i Mans Minde, Folkminder øste af Folkemunde i det 19 Aarhundred,* Odense, R. Hansen, 847 p.

RYDÉN, Reine (éd.) (2004), *Jordbrukarnas kooperativa föreningar och intresseorganisationer i ett historiskt perspektiv,* Stockholm, Skogs- och lantbrukshistoriska meddelanden nr 32.

SÄVE, Gunnar (1914-1917), *Jönköpings läns Kgl. Hushållningssällskaps historia 1814-1913,* Jönköping (Sweden), 1, 572 p., 2, 186 p., 3, 414 p. http://libris.kb.se/bib/1241941?vw=full

SOMMESTAD, Lena (1992a), 'Able Dairymaids and Proficient Dairymen: Education and de-Feminization in the Swedish Dairy Industry', *Gender & History,* 4, 1, p. 34-48.

SOMMESTAD, Lena (1992b), *Från Mejerska till mejerist, En studie av mejeriyrkets maskuliniseringsprocess,* Lund, published dissertation Uppsala University..

SVENDSEN, Gunnar L.H., SVENDSEN, Gert T. (2000), 'Measuring Social Capital: The Danish Cooperative Movement', *Sociologica Ruralis,* 40, 1, p. 72-86.

VAN ZANDEN, Jan Luiten (1991), 'The first Green Revolution: the Growth of Production and Productivity in European Agriculture, 1870-1914', *Economic History Review,* 1991, II, p. 215-239.

WINSTANLEY, Michael (1996), 'Industrialization and the Small Farm: Family and Household Economy in Nineteenth-Century Lancashire', *Past and Present,* 152, p. 157-195.

WRIGLEY, Edward A. (1988), *Continuity, Chance & Change. The Character of the Industrial Revolution in England,* Cambridge, Cambridge University Press.

12. The iron cage of dairy farming. Self-sufficiency and specialisation in Finnish peasant farming at the beginning of the twentieth century

Matti PELTONEN

I. Introduction

How did the small-scale family farm manage to survive in the conditions of an underdeveloped market? There are two standard answers. One claims that abundant family labour explains its inherent strength; this was the view of populist Marxists and the followers of Chayanov. The other answer is that it had to struggle against decreasing labour productivity and that this caused income to fall and eventually become negative; that was the outlook of orthodox Marxists and also of the Classical School of economics who saw the situation from the point of view of large-scale farming, and presented figures to demonstrate that income rose for each unit of size in the farming unit. Those who defended the future of small-scale family farming and the co-operative movement which was part of their credo, pointed to the higher yields per unit of surface on smaller farms. These stereotypical comparisons of small-scale and large-scale farming, however, neglect many aspects of traditional peasant farming in Western Europe. Exploring such matters requires looking into the unequal growth of product commercialisation in different sectors of agricultural production. It also requires us to examine the uneven development of different productive assets such as arable land, buildings, work-force and so on.

This paper attempts a fresh look at these old debates on the fate of family farming in competitive market conditions. It is based on an empirical examination of how the market economy made its impact on different types of small-scale farming (freeholders and tenant farmers) in a peripheral region of Europe. Market impact is measured by the mode and degree of commercialisation and specialisation in agricultural production. In my dissertation, I already made some use of this data, concerning the economic conditions of about 1,200 freeholders and nearly 600 tenant

Agricultural specialisation and rural patterns of development, ed. by Annie ANTOINE, Turnhout, 2016 (*Rural History in Europe*, 12), p. 233-247.

BREPOLS PUBLISHERS DOI 10.1484/M.RURHE-EB.5.112270

farmers in 1903[1]. Initially, I was interested in the degree of commercialisation, which was not very high on average. Surprisingly, on smaller farms (those with less than 10 hectares of arable land) of equal size, the marketed share of all agricultural production seems to have been the same on both freeholds and tenant farms at the beginning of the twentieth century: about 36 per cent on freeholds and about 30 per cent on tenant farms. But this analysis can be pushed farther in looking at the question from the point of view of specialisation and what it meant to farm households of different types and in different regions.

Table 12.1. Share of the marketed production on independent and tenant households in five Finnish parishes, 1903

	Arable per farm in hectares						
in percentage	*< 2*	*2-5*	*5-10*	*10-15*	*15-25*	*25-100*	*> 100*
Independent	29	32	35	44	46	58	77
Tenant	29	28	33	32	30	43	-

Note. Value of sold timber included.

Source. PELTONEN (1992: 225).

Finland took part in the agricultural world market of the late nineteenth and early twentieth centuries as an importer of bread-grains, since the country was only about 40 per cent self-sufficient in these cereals before World War I. Customs protection for agriculture extended only to dairy products: the production and import of margarine was prohibited. The Finnish Senate subsidised dairy loans and supported the shipment of butter to the main export harbour of Hanko by providing special refrigerated wagons and reduced shipping fees for export products. The state also provided ice-breaker services during the harshest winter months to make sure that Finnish butter and forest products could reach the Continent and Great Britain.

Producer households in Finnish agriculture were of two sorts. The first were the landowner households – so-called 'independent' farms – usually occupying

[1] A short summary of my dissertation is given in Peltonen (1993). The data used in this paper was originally collected in 1904 by an official committee appointed by the Finnish Senate in order to determine the level of income of different social strata in the countryside. The data was used in calculating insurance fees and compensation for the incapacitated. The 1903 data covers practically all farming households with more than 2 hectares of arable land in five parishes (Tuusula, Humppila, Kymi, Räisälä and Vihanti). The location of these parishes is indicated in Figure12.1 at the end of this article. Part of the material was published at the beginning of the twentieth century as GYLLIN G. (1907), I am, however, using the unpublished part of the data collected to determine the overall net income of rural households.

Figure 12.1. Territorial division of Finland by counties

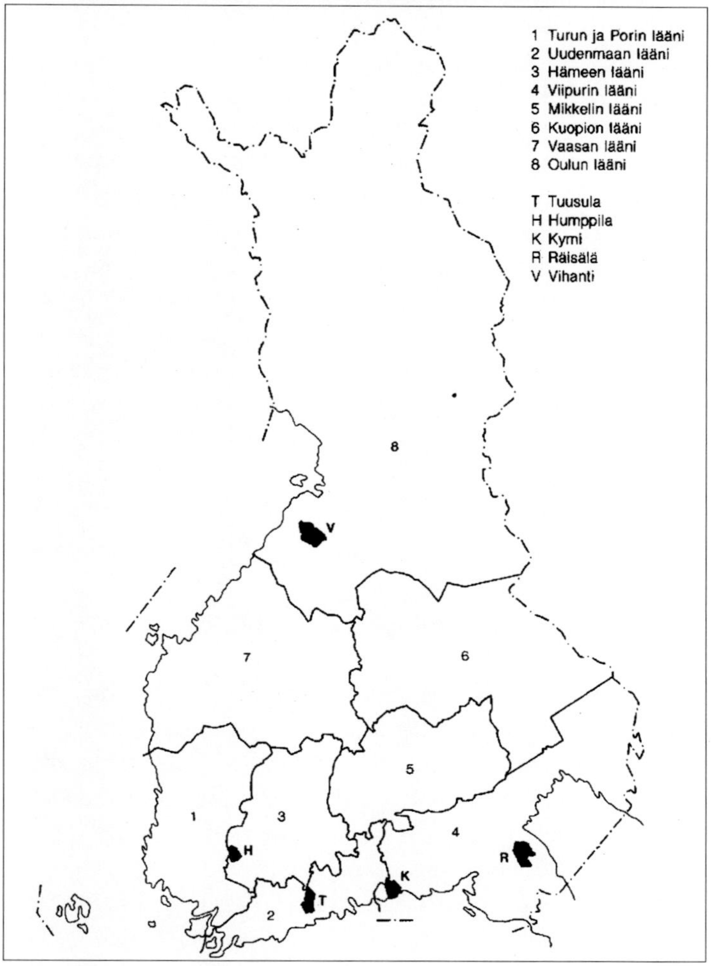

Source. Map by the author.

from 2 to 25 hectares of arable land. In 1910 there were a little over 100,000 of these 'independent' farms, but only a tiny fraction of them were manor houses with more than 100 hectares of arable. It was reckoned that landowners with more than 25 hectares of arable land usually spent their time running the farm and no longer took part in actual agricultural work. Accordingly, such farms were more apt to use wage labour. The other kind of producer household consisted of small tenant farmers, called torppari (roughly the same term as 'crofters' in English), with usually 2 to 15 hectares of arable land. In 1910 there were almost 43,000 such farms. Tenant farmers with fewer than 2 or 3 hectares, usually called cottagers (mäkitupalainen in Finland or backstugusittare in Sweden) were even more numerous[2]. They were considered to be distinct from crofters because they were believed to get most of their income from work outside the household. In addition, there was a large group of farm servants and worker families living in their masters' quarters as 'lodgers'.

The productivity of the main sectors of the farm economy was low compared to other Nordic countries (especially Denmark and Sweden). This was despite measures taken, even on the smaller farms, to apply deep ploughing with iron ploughs, to grow hay from seed, and to set up small 'village dairies' throughout the land. All these policies were effectively implemented in the 1870s. But the obligation on tenant households to pay their land rent in labour service exercised a stranglehold on their agricultural production and prevented them from earning extra income from activities outside the farm[3]. The overwhelming majority of landowners receiving land rent in the form of labour services were peasant freeholder households. These freeholder households also employed, to some degree, non-family labour, in traditional employments as servants in husbandry or as agricultural day-labourers, both groups being partly remunerated in cash.

II. Forms of self-sufficiency on the farm level

My aim is to present data on the relative importance of the various sources of cash income, and to assess how different types of agricultural households could engage in different kinds of specialisation. The results will show that all Finnish farms relied heavily on dairy farming, especially tenant farms. The latter however were severely hindered from supplying their produce to the market – especially beyond the local market – because of their institutional position. These enjoyed a high degree of specialisation: 68 per cent of cash income on freeholds and 90 per cent on tenant farms came from animal farming, mainly fresh milk and butter (Table 12.3). But this specialisation was not the sign of a spectacular success story.

Usually the Finnish case is analysed using the two-sector model familiar from traditional development economics, dividing the economy into a modern, dynamic and growing sector, export industries, and a traditional, stagnant sector, agriculture, with indefinite labour resources. In this model, agriculture flourished only by selling timber to different branches of the forest industry. My differing views on this question have been influenced by the Polish economic historian Witold Kula's structural

2 Tenant farms of less than 2 hectares of arable land numbered 53,234 in 1910.

3 Modern forms of land-lease contracts were not common in Finland before the First World War; most of the agricultural land was leased by crofters (*Fi. torppari*) and cottagers (*Fi. mäkitupalainen*) and the land rent was paid for the most part by male labour services, with or without horses.

analysis of Polish agriculture in the early modern period. Kula's book contains a good criticism of the two-sector model, which I shall attempt to prolong by applying his points to a different period (Kula, 1976: 20-23)[4]. The starting point is that 'modern' and 'traditional' meet on the farm level. These are not clear-cut economic segments, but overlapping and intersecting sectors causing contradictions and economic difficulties for the households. There is a large body of research by scholars interested in the special features of only partly commercialised peasant households, where these conflicts and contradictions are explained away by concepts like 'the consumption orientation of family-farming', 'simple versus extended commodity production', or 'articulation of the modes of production'.

In crop production two remarkable features stand out (see Table 12.2). First, very little of domestic crops (grains, potatoes or hay) was put directly into the market. Secondly a great deal of the value of crops produced was accounted for by fodder crops, mostly hay. Of course, the price level of hay used to estimate value is a bit unreliable because hay occupied a small share of marketed production and did not have to compete with imported fodder, unlike most other crops. The degree of self-sufficiency in bread grains (rye and barley) is also striking. Along with the low overall level of national self-sufficiency in bread grains, it indicates that a large part of the rural population -- the small tenant households and worker households -- relied on rye flour imported from abroad. This stubborn persistence in making their own bread from the produce of their own fields had two quite different reasons. On the one hand, all farming households were short of cash and, on the other, they preferred bread from home-grown cereals for sentimental reasons. Moreover rye bread made of grains dried in the Finnish style, in a *riihi* (where the grain sheaves were in contact with smoke before the actual threshing), had a different taste than bread made from imported flour. This attachment to growing bread grains also kept smaller farms from adopting more advanced crop rotations. Nearly all Finnish farms were also self-sufficient in fodder for their cattle and horses. Bread grains and potatoes were also used as animal fodder on Finnish farms during the decades before World War I. But there was no automatic guarantee that the labour expended in growing fodder crops was fully compensated by, for instance, lower prices for milk or butter[5].

---

4 Kula makes three criticisms of the two-sector model. First, the border between the traditional and modern economy can not be drawn between enterprises. Secondly, even if the traditional sector had an unlimited reservoir of labour, to mobilisation it out of the traditional sector would have required extraordinary measures. Thirdly, to suppose complete stagnation in the traditional sector is not reasonable.

5 It has been stressed, for instance by Witold Kula, that when a natural economy prevails, enterprises are either clearly unprofitable, if one transforms all the costs involved into monetary terms – or quite profitable, if one takes into account only the costs that are expressed in money terms. See KULA (1976: 35) and MAGNUSSON (1980: 126-127).

Table 12.2. Crop production, animal farming and forestry in five Finnish communities, 1903

Product (a)	*Total value of production*		*The use of the product*					
	Independent/	*Tenant*	*Independent*			*Tenant*		
			Con-sump-tion	Foder, seed	Sale	Con-sump-tion	Fodder, seed	Sale
Crop production:								
Wheat	0.1	0.3	80	16	4	93	7	-
Rye	15.3	14.5	75	23	2	74	25	1
Oats	4.5	5.9	48	50	2	61	38	1
Barley	15.6	14.6	4	92	4	7	85	8
Potatoes	8.9	11.1	62	30	8	66	28	6
Hay	54.9	52.2	0	95	5	0	98	2
Other	0.7	1.4	87	11	2	86	13	1
TOTAL	100.0	100.0	23	71	6	28	69	3
Animal farming								
Milk, cream	56.3	44.1	47	-	53	71	-	29
Butter	22.7	30.1	42	-	58	42	-	58
Beef and veal	7.2	10.7	35	-	66	23	-	76
Other	13.8	14.3	76	-	24	78	-	22
TOTAL	100.0	100.0	49	-	51	63	-	37
Forestry								
TOTAL	100.0	100.0	56	-	44	95	-	5

Note. (a) Share of the total value of agricultural production and its use on independent landowner and tenant farmer households.

Source. PELTONEN (2004: 85).

When we look more closely at the relative share of the various crops in overall production, we can see that the division of the farm economy into commercial and self-sufficient parts is not as straightforward as numerical criteria, such as the 50 per cent threshold of commercialisation, would suggest. The marketed share of production, especially marketed dairy products, depends totally on the amount of self-sufficient fodder production on the farm. And, although they were highly unprofitable, bread grain crops continued to be grown on the farm, because cash income was low, owing to the peculiar conditions of Finnish farming and its relationship to international competition on agricultural markets. Hay growing was however the most mechanised part of crop production on Finnish farms at the beginning of the twentieth century.

Horse-drawn mowers and rakes were used in harvesting hay from the arable land, while the bread-grain harvest was done with traditional sickles in order to secure the maximum yield.

Animal farming, taken all together, made up the commercial part of Finnish farming at the beginning of the twentieth century. This comes as no surprise: traditionally this period of Finnish agriculture is called 'the farming system dominated by dairying' (Simonen, 1949). It is astonishing to find that beef and veal were the most commercialized items of marketed production. The amount of meat produced was, however, not very significant compared to total milk products (fresh milk, home-made butter and cheese). Nearly all rural households were eager to sell as much beef and veal as they could, a sign of how desperate they were to earn every last possible bit of cash income.

Forestry seems to have been the most self-sufficient part of the Finnish farm economy at the beginning of the twentieth century. This is another surprise, because traditionally, forest income was seen as having played a strategically and economically important role in the commercialisation and modernisation of the Finnish countryside[6].

There were several reasons why the income from selling timber could not be as high as was thought in the traditional view of Finnish agricultural history. First, only a small fraction of landowner households had enough forest to sell their product. An expert committee, analysing the situation of privately owned forests at the end of the 1890s, calculated that only about one third of all landowners were in a position to sell their timber. The second reason was the unfavourable situation of producers in the market for raw wood, where a few buyers faced thousands of sellers. For many decades landowners could exert no positive influence on the prices they obtained, and needed protection from the state and their own organisations.

Forest products were used in the farming household for building houses and wooden fences, and for heating the living quarters, the sauna and the *riihi*. Moreover chopped fir branches were used as litter for the cows when they were fed on stored fodder during the winter months. This wood material ended up as manure on the arable land. The nineteenth century witnessed a remarkable rationalisation of timber use. This was done by using more economical heating systems and improving the construction of houses and other buildings by such devices as building multipurpose buildings and putting a stone layer between the ground and the wooden parts of the house. It has

6 See for example, the studies of Eino Jutikkala and K. O. Alho, published in the 1940s as representative of this line of thought about the special importance of income from selling timber from the peasant farms. Cf. JUTIKKALA (1946); ALHO (1949).

been claimed that the most important change during the nineteenth century in forestry occurred on the farm and not in the market (Soininen, 1974: 292-300).

We have already mentioned, at the start of this paper, the traditional system of land-lease prevalent in Finland at the beginning of the twentieth century. Landlords drew on a pool of agricultural labour formed by their tenants' families and their cottagers and lodgers; this was another form of rural self-sufficiency and a convenient way to avoid hiring labour for cash. The relationship of landowners to their underlings was different in different parts of the country. In southwestern Finland, in the three agriculturally most advanced provinces, about two thirds of all landowners had tenants, either crofters or cottagers, on their lands. In southwestern Finland there were on average 4-5 families of tenant households for one landowner, while in eastern and northern Finland the few existing landowners had only an average of 2-3 tenant families each.

III. The importance of dairy farming

In 1903, almost all cash income on Finnish farms came from dairy farming, mainly from selling fresh milk or milk products such as home-made butter and occasionally home-made cheese; only to a lesser degree did it come from selling timber from freeholds. As with all rural households, there was also another main potential source of cash income: the sale of labour power, either wage-labour or horse transportation services; this secondary income from outside the farm economy is not included in the sales income detailed in Table 12.3.

The market for dairy products was clearly divided into two segments in the late nineteenth and early twentieth centuries. Most of the butter and cheese produced in modern dairy plants was exported abroad. The price of export butter was set in Copenhagen and milk producers were paid for their produce according to the weekly price bulletins. Domestic demand for butter was met for the most part by home-made butter; prices for this product did not necessarily fluctuate at the same rate as export prices. Home-made butter was also a more local product and did not travel such time-consuming distances as export butter.

Specialisation in dairy farming was an 'iron cage', because it seemed to be the only viable option open even to small-scale farmers, unless they accepted the alternative of engaging in full-time wage-labour. When competition in the world market for dairy products grew in the early years of the twentieth century, specialisation in dairy farming became even more difficult, and often unprofitable. In terms of the marketed share of farm production it was the small tenant households that were more fully specialised in dairy farming. Households owning land, especially the smaller ones, could also sell various forest products such as firewood, timber, hand-hewn timber, tar and charcoal. Small farms appear in the statistics to be selling a proportionately larger share of wood products; this is because their revenue from this item concealed a large amount of labour income. In order to earn more cash from selling timber, landowners felled and transported some of the forest products and sometimes took part in the finishing stages, for example in the preparation of firewood.

Table 12.3. Distribution of the value of sales production in five Finnish parishes, 1903

	Arable land per farm in hectares						
in %	*-2*	*2-5*	*5-10*	*10-15*	*15-25*	*25-100*	*100-*
Independent farms							
Crop production	3	8	6	10	7	13	13
Animal farming	53	62	65	72	77	81	62
Forestry	34	27	27	17	15	6	25
Other	10	3	2	2	1	0	0
Total	100	100	100	100	100	100	100
Tenant farms							
Crop production	6	6	9	7	14	10	-
Animal farming	92	89	91	93	86	90	-
Forestry	2	5	0	0	0	0	-
Other	0	0	0	0	0	0	-
Total	100	100	100	100	100	100	-

Comments. The figures exclude secondary earnings from working outside the farm, but the income from forestry includes work income, because at this time a considerable part of the sold timber was felled and transported to the nearest water-way or railway station by the seller.

Source. Peltonen (1992: 234–235).

Although tenant farmers were more specialised in selling dairy products, they worked under several handicaps. Tenant farmers, as milk producers, were usually allowed to take part in running modern dairy establishments. But it was often hard for them to produce enough milk every day for delivery to the dairy. Crofters, whose farms were usually situated on the outskirts of the village, frequently had to move their products over greater distances. Tenant farmers often had more trouble than independent landowners in providing both summer pasture and winter feed for the cows. It is quite clear (see Table 12.4) that tenant farmers chose to turn their milk into home-made butter more frequently than independent farmers. This butter was sold to the village shopkeeper or in the nearest town marketplace; there was no retail market for fresh milk in the countryside, because every family tried to produce its own. There were also great regional differences in the structure of dairy farming. In the north of the country there were no towns or other population centres such as industrial complexes and army garrisons, and it was harder to operate a modern dairy because only some of the milk producers could be counted on regularly.

Table 12.4. Fresh milk in the value of dairy products in five Finnish parishes, 1903

(a)	*Independent farms (%)*	*Tenant households (%)*
Tuusula	93	63
Humppila	82	71
Kymi	95	100
Räisälä	30	6
Vihanti	5	1

Note. (a) Share of the value on independent and tenant households.

Source. PELTONEN (1992: 165). For the location of the five parishes, see Figure 12.1.

At the beginning of the twentieth century Finnish tenant households, and also the bigger independent landowners[7] combined a low level of commercialisation with a high degree of specialisation. As we have seen, the situation was not the same with independent and tenant farmers because of differences in their institutional positions.

IV. Immovable stocks

Time is an important factor in agriculture, not only in seasonal crop production, or in phases of the family cycle, but also in other basic economic processes of the farm economy. In a predominantly non-monetary economy the problem of capital is crucial. The treatment of capital was already considered in the 1970s to be the weak point in the peasant-economy model put forward by scholars following in the steps of Chayanov (Peltonen, 1984). Russian historian V. P. Danilov proposed a new definition of the borderline between peasants and farmers. For him 'natural' productive forces in the peasant economy like land, cattle, seed and manure are more important than 'historical' productive forces such as tools and machinery. The conditions of agricultural production in some regions in imperial Russia might possibly be

7 In this regard, all independent landowner households with more than 10 hectares of arable land might be described as specialising in dairy farming.

described in these terms. But usually 'land', like other 'natural' productive assets, is a very historical factor indeed. To cultivate a plot requires the work of generations in reclaiming the land, clearing the stones, digging drainage ditches, manuring and otherwise improving the soil and constantly making the field ready for sowing. All this demanding work and investment in kind is reflected in land value. Even in peasant agriculture, the value of land outstripped that of other assets such as tools, cattle and buildings[8].

As the banking sector, and especially land mortgage banks, developed, people became more aware of land values, even on the smaller farms[9].

But this is not all. One difficulty in peasant agriculture was the different time dimensionality of economic phenomena: the difference between stocks and flows. Income flows were only partly in money; landowners in particular, but also other producer households, got much of their income in kind (see Table 12.5). Yet productive assets, and agricultural land most of all, had to be completely transformed into money in order to pass from one owner to the next. During this period, it became very hard to get through certain points in the family cycle, such as the transmission of the farm and farm assets from one generation to the next. Even if financial institutions such as commercial banks, mortgage banks and saving banks gave long-term credit to farmers, the low level of money income created intolerable problems for farm management. As an old Finnish proverb put it: 'If you have no money, you have no money, even if the farm is big'. All these contradictions and difficulties came together in acute form when land prices shifted during the decades of the Grain Crisis, which, in Finland, lasted from the 1880s down to World War I. Land prices remained high, and even rose despite low or non-existent profits in agriculture. To make matters worse, the transfer fees paid by each new tenant household to its predecessor were already astonishingly high by the end of the nineteenth century.

Usually this phenomenon is explained by the dual economy model, and the rise of forestry and the forest-based export industry are made to account for the rising land prices (Jutikkala, 1946)[10]. However, one significant reason why land prices rose in addition to heavy investment to soil improvement, was that there was no longer a difference between the price at which landed property sold to family members and

[8] The problem caused by neglecting capital (or even property) results of course, from one of the axiomatic assumptions of the 'dual economy' theory, where agriculture is supposed to use only two factors of production, land and labour (and industry, capital and labour).

[9] In Finland the first land mortgage bank, Suomen Hypoteekkiyhdistys, was established at the beginning of the 1860s.

[10] It should be noted however that the prices of different kinds of wood products did not rise in any noticeable way at the end of the nineteenth century.

relatives, and to people from outside the family (Peltonen, 2001: 54-66). For several decades farms transmitted within the family had been evaluated at unchanging prices. Although the commercialisation of agriculture and the level monetary income remained quite low, the new rise in land prices did create problems for landowner families if for example they to purchase land on the open market. The money share of their income could be small and they were eager to get extra revenue from working outside the farm, one way or another.

V. The dark side of specialisation

We already saw how small independent farmers compensated for their low revenues from dairy farming by selling timber and using labour-intensive arrangements in their selling practices. By contrast, tenant farmers could not, as a rule, sell forest products at all. There were, however, some exceptions to this rule. It seems that the rules for forest use were in reality more lax in the north of the country, perhaps because the state owned a large share of the northern forest, and so tenant farmers took timber from those woods more freely. The other option for tenant farmers was to work outside the farm. Secondary income from outside the farm was an important part of the cash income for tenant households, amounting to almost half of all their cash revenue in the 1903 statistics. On independent farms the share of secondary income was also noticeable, amounting to almost one-fifth.

Table 12. 5. Secondary income in five Finnish parishes, 1903

in percentage	*Independent*	*Tenant*
Frequency of secondary income	62	57
Share in total income	9	21
Share in cash Income	18	48

Source. Peltonen (1992: 203-205).

Secondary income came mostly from providing horse transportation services or wage labour. Independent farm households got nearly all their secondary income from selling horse transportation to other sectors of the economy: industry, forestry, retail trades and construction. For tenant farmers, the sale of horse transportation was also the biggest source of secondary income, followed by wage labour or handicraft production. Unfortunately it was only during the winter months that farmers could easily work away from the farm, whereas it was in summer that the best opportunities for outside work could be found just at the time when tenant households had to furnish labour to landlords as part of their rent.

As a result of all this, incomes and consequently the level of well-being on small tenant farms were not very satisfactory. Even after a long year of hard work the resulting income often came to little. The level of income clearly reflected the resources available to tenants and their place in the social hierarchy.

Table 12.6. Average income and share of monetary income in different agricultural occupations in five Finnish parishes, 1903

	MALE			*FEMALE*		
	N	*Income*	*In cash, %*	*N*	*Income*	*In cash, %*
Landowner	1,248	1,372	45.4	110	1,358	58.9
Tenant	542	690	34.9	46	539	37.9
Agr. servant	356	408	40.7	397	301	32.8
Agr. worker	537	361	77.7	125	193	68.4
Adult family m.	1,252	264	12.0	877	161	4.6

Note. According the gender of the head of the household.

Source. GYLLING (1907, 92-93). The non-monetary part of income includes the value of food, clothing and housing. Some of the 'households' comprise only one person (servants, adult family members). These figures are for adult family members other than the married couple or elderly person on a pension. They include only families whose members worked in agriculture.

Beside the tenant farmers there was yet another underprivileged group in the agricultural rural population: adult family members other than the married couple living in landowner and tenant households. Unfortunately, in the data analysed in this paper the adult members of rural households, single and married, were all lumped together regardless of the kind of household to which they belonged. Despite this statistical defect, they clearly earned the least among the rural work force, receiving even less than agricultural servants. Edward Gylling, chief statistician of the so-called Invalidity Committee, suggested as much when he remarked, in commenting upon its findings at the beginning of the twentieth century, that 'The lowest income level are those whose children work at home, and by living at home do not receive full compensation for their work input' ('Alimmat tulot on kotona työskentelevillä lapsilla, jotka kotona asuen eivät saa täyttä korvausta työstänsä') (Gylling 1907: 93). The Committee's statistics showed that family members working in agriculture were young, more than 80 per cent of them being between 16 and 30 years of age. Most of them were unmarried adult family members (75 per cent of the males, and 90 per cent of females). The bulk of family members working in agriculture (80 per cent of the males and almost 90 per cent of the females) did not receive any cash income[11].

VI. Conclusion

The conventional distinction between modern and traditional sectors of the economy needs to be abandoned, for the borders between modern and traditional phenomena are not located neatly outside the enterpreneurial sector. Rather the contrary: they are situated within the enterprises. In a complicated enterprise like the peasant farm, modern and traditional aspects meet at many points and create new divisions inside these units. In this paper I have proposed that inconsistent price structures, the undervaluation of uncommodified labour and the different dimensions of stocks and flows in important economic categories are issues in need of more careful examination.

Instead of trying to characterise early twentieth-century peasant households by one single characteristic such as peasant farming, family farming or simple commodity production, I have tried to find areas where 'old' and 'new' phenomena meet and cause friction. One such area discussed here was the choice between selling fresh milk to the dairy or to consumers living nearby, or retailing it as home-made butter. Another concerned allocating farm resources to fodder production, so as to promote dairy farming, or to bread grains, to preserve self-sufficiency. Many landowners also had to opt for using wage-labour or the labour of their family members and tenant farmers. And, especially in the eastern parts of the country many small farmers had to decide between concentrating on farming or on securing secondary earnings.

Bibliography

Alho, K. O. (1949), *Suomen uudenaikaisen teollisuuden synty ja kehitys 1860–1914,* Helsinki, the Bank of Finland.

Figes, Orlando (1988), 'V. P. Danilov on the Analytical Distinction between Farmers and Peasants', in T. Shanin (ed.), *Peasants and Peasant Societies: Selected Readings*, 2nd edition, Harmondsworth, Penguin, p. 121-124.

Gylling, Edvard (1907), *Väestö-, invaliditeetti- ja tulosuhteet Tuusulan, Humppilan, Kymin, Räisälän ja Vihannin kunnissa vuonna 1903. Tilastollinen tutkimus*, Helsinki, Invaliditeettivakuutuskomitea.

Jutikkala, Eino (1946), *Suomen talonpoika kautta aikojen,* Porvoo, Werner Söderström.

[11] Most married family members were found in the parishes of Kymi and Räisälä in Eastern Finland, where family members as a rule were also more numerous than elsewhere in the data.

KULA, Witold (1976), *An Economic Theory of the Feudal System. Towards a model of the Polish Economy 1500-1800,* transl. by Lawrence Garner, London, Verso.

MAGNUSSON, Lars (1980), *Ty som ingenting angelägnare är än mina bönders conservation…,* Godsekonomi i Östra Mellansverige, Uppsala, University of Uppsala.

PELTONEN, Matti (1984), 'A. V. Tshajanovin teoria talonpoikaistaloudesta', *Sosiologia,* 21, p. 298-308.

PELTONEN, Matti (1992), *Talolliset ja torpparit. Vuosisadan vaihteen maatalouskysymys Suomessa,* Helsinki, Finnish Historical Society [English summary: *Landowners and Crofters. The Peasant Question in Finland at the Turn of the Century*, p. 399-429].

PELTONEN, Matti (1993), 'The Peasant Economy and the World Market: Finnish Peasant Farming in the Age of Agrarian Crisis, 1880s-1910s', *Review Fernand Braudel Center,* 16, p. 357-381.

PELTONEN, Matti (1998), 'Från osäkerhet till hat. Torparfrågans moraliska ekonomi i sekelskiftets Finland', *Bebyggelsehistorisk tidskrift,* 35, p. 91-98.

PELTONEN, Matti (2001), 'Maahengen katoaminen ja maalaisväestön mentaliteettimuutos 1800-luvun lopulla', in Jukka KORPELA, Tapio HÄMYNEN & Arto NEVALA (eds), *Ihmisiä, ilmiöitä ja rakenteita historian virrassa,* Joensuu, University of Joensuu, p. 54-66.

PELTONEN, Matti (2004), 'Uudet kaupallistumisen muodot', in Matti PELTONEN (ed.), *Kasvun ja kriisien aika 1870-luvulta 1950-luvulle: Suomen maatalouden historia 2,* Helsinki, Finnish Literature Society, p. 77-134.

SIMONEN, Seppo (1949), *Lypsykarjatalousvaltainen tuotantojärjestelmä Suomessa: Maataloutta ja maatalouspolitiikkaa vuosisadan vaihteen molemmin puolin,* Porvoo & Helsinki, Werner Söderström.

SOININEN, Arvo M. (1974), *Vanha maataloutemme. Maatalous ja maatalousväestö Suomessa perinnäisen maatalouden loppukaudella 1720-luvulta 1870-luvulle,* Helsinki, Finnish Historical Society [English summary: *Old Traditional Agriculture in Finland in the 18th and 19th Centuries*, p. 448-459].

13. Wine-growing and agricultural specialisation in Late Medieval and Early Modern Europe

Erich LANDSTEINER

I. Introduction

During the last twenty-five years the focus in the study of the economic development of European regions has clearly shifted from technology, productivity, demography and social structure to the role of market development, market integration, and transaction costs (e.g., De Vries, 1994; Grantham, 1999). In the light of this changed emphasis, wine-growing and wine marketing are of considerable interest. 'No other form of agriculture was so commercialized' (Brennan, 1997: XVII). Wine has been called the agricultural 'market product par excellence' (by Charles Tilly, cited in Brennan, 1997: XVI) of medieval and early modern Europe. The sceptical argument, less frequently encountered in France then elsewhere, that wine could have been only of minor importance in an economy based essentially on cereal production is misleading insofar as wine production was much more widespread, far into the early modern period than nowadays. As large quantities of grain were consumed directly by the producers and an important part of the harvest had to be set aside for seed, the value of wine production in some parts of Europe could have well equalled the value of the grain harvest that actually entered the market. Even if we calculate total production in monetary terms the share of wine remains impressive, as can be seen from Table 13.1 which provides data for France and Lower Austria in the middle of the nineteenth century. Today (2004), wine accounts for 5.3 per cent of the EU's agricultural output – almost the same as the share of wheat, 6.5 per cent (European Commission, 2006: 28).

In relation to the subject of this paper, it makes sense to distinguish between three levels of specialisation: the plot, the individual agricultural holding, and the region. In what follows, it will be shown that on all three levels wine-growing in the European

Agricultural specialisation and rural patterns of development, ed. by Annie ANTOINE, Turnhout, 2016 (*Rural History in Europe*, 12), p. 249-271.

BREPOLS PUBLISHERS DOI 10.1484/M.RURHE-EB.5.112271

past was not necessarily specialised, that its position in the agrarian economy depended very much on the agrosystem in question, and that access to markets and transport facilities played a major role in opening or closing the road to specialisation. Wine possessed some destinctive features as a commodity in an 'ancien régime-economy' (Grenier, 1996: 9-15), notably its high price, its income elasticity of demand and substitutability. As a result, not only the accessibility and extent of the potential market but also its depth made a big difference to the viability of specialised wine production. Shrinking mass purchasing power in periods of dearth and immensely fluctuating yields and price volatility, especially in northern wine-growing regions, made for market instability. In such an economic environment specialised wine production was a very risky business, and to counterbalance these risks producers had to develop strategies appropriate to their own location, resources, and social position. Finally, in the past as well as in the present, wine is not a homogenous commodity. Its markets are therefore highly segmented, and producers and traders have to deal with demand behaviour that varies according to its quality, prestige and price.

Table 13.1. Wine-growing area and value of wine production in France, 1852 and Lower Austria, 1859

	(a)		*(b)*			
	Area (%)		*Gross value of production (%)*		*(b)/(a)*	
	France	*Lower Austria*	*France*	*Lower Austria*	*France*	*Lower Austria*
Cereals	44.60		57.28	60.55	1.28	
All arable	76.05	72.79	78.28	70.41	1.03	0.97
Wine	6.36	4.13	9.05	11.65	1.42	2.82
Grassland	14.68	23.08	10.81	17.94	0.74	0.78
Treecrops	9.27		10.65		1.15	
Total	100,00	100.00	100.00	100.00		

	Absolute values for wine	
	France 1852	*Lower Austria 1859*
Area (ha)	2,191,000	45,600
Production (hl)	38,090,400	499,500
hl/ha	17.38	10.95
	(79% of 'average yield')	(60% of 'average yield')

Sources. DEMONET (1990:130, 184); LORENZ (1866: 292-293).

These problems will be discussed in detail on the basis of evidence from different European regions and the long history of wine-growing in this part of the world, although the uneven nature of the existing literature necessarily makes coverage patchy. But first, let us take a short but quite revealing look at the present level of specialisation in EU wine production.

In February 2006, the Directorate-General for Agriculture and Rural Development of the European Commission published an extensive working paper on the economic state of the EU wine sector. Here, specialised wine-producing holdings are defined as 'holdings receiving at least 2/3 of their gross farm income from the production of wine and associated products' (European Commission, 2006: 91, n. 37). The distribution of specialised holdings clearly displays the divergence between the Mediterranean and the more northern or continental European countries. This difference in specialisation correlates with the proportions of table wine and quality wine production in eachcountry. The definition of the respective areas of table wine and quality wine production is a matter for each member state, which makes comparison tricky, but it is nevertheless evident that specialisation goes along with quality wine production. This is also the conclusion drawn in the report of the European Commission: 'Specialist wine-producing holdings represent the majority of quality wine holdings and more than three-quarters of the land in that category; by contrast, for more than three-quarters of the holdings producing table wine that activity is a secondary one' (European Commission, 2006: 102). Wine-growing holdings in Greece and Portugal have only very small parts of their holdings planted in vines (0.6 ha in Greece and 0.9 ha in Portugal), whereas the average French wine-growing holding contains 7 ha planted with vines (European Commission, 2006: 100-102).

Although one might think that these differences are the product of recent developments, I want to argue that these data reflect a long-standing fundamental difference in the position of wine production within the agricultural economy of the Mediterranean and the extra-Mediterranean parts of Europe, which lingers on until today.

II. Mediterranean and extra-Mediterranean viticulture, or extensive and intensive wine-growing

In speaking of the history of wine production and consumption, it is common to divide Europe into three zones (Dion, 1959: 4; Braudel-Spooner, 1967: 407-411):

– the Mediterranean area, the boundary of which is marked by the northernmost cultivation of the olive tree. Here vine growing encountered no climatic obstacles and was all-pervasive. Wine was therefore part of the daily popular diet;

– an intermediate zone where the plant had to be adapted to an adverse climate, which had to be mastered by the tenacity of the growers, who had to expect much higher production costs and more unreliable results than in the Mediterranean area;

– a northern area whose climate makes commercial viticulture impossible. Although this zone constituted an important market for the more southern regions of production, wine here was more or less a luxury good confined to the upper classes and special occasions.

One of the defining features of the Mediterranean agrarian ecosystem is the frequent intercropping of cereals, tree crops and vines. Because of its integration into Mediterranean agrosystems, wine-growing was nearly ubiquitous in this part of Europe. The exchange of wine for other produce took place, if at all, mainly on the local level. Nearly every town and village in the Christian Mediterranean produced its own wine, and most landowners had at least a small parcel planted in vines, often only for their own needs (Braudel, 1966: I, 215-217; Horden-Purcell 2000, 175-230).

One might object that there are quite prominent examples of inter-regional wine trade in Mediterranean history. What about the sweet wines of the Aegean islands, which played a major role in medieval long-distance trade? What about Andalusian Sherry, Sicilian Marsala or the famous Port from the Portuguese Douro valley? What about the export of wine and brandy from Catalonia to the colonies of the Spanish Empire in the eighteenth and nineteenth centuries, which led to monoculture by way of commercial specialisation (Vilar, 1962: II, 322-327)? These prominent exceptions confirm the rule! They are either based on the production of special wines for distant markets or they are connected to the penetration of either colonial production zones or colonial markets.

The general contrast in the degree of specialisation and market integration between Mediterranean and extra-Mediterranean viticulture is neatly confirmed by the way vines were raised in the two macro-regions. The difference between the two modes of cultivation was so drastic that one author wrote about two kinds of 'viticultural civilisations' with respect to France and Italy (Loubere, 1978: 84). It is probably more appropriate to speak of extensive and intensive forms of wine-growing. In the north monocultural vineyards predominated, in which the plot was planted only with vines and possibly one or two fruit trees. The wines were trimmed low and the shoots were fastened to wooden stakes or more complicated constructions for support. The density of plants was high: 20,000-25,000 vines per hectare were quite common (Durand, 1979: 314; Landsteiner, 1992: I, 45; Laurent, 1957: I, 181; Musset, 2008: 177). This high planting density and the practice of renewing or supplanting exhausted plants by burying them or their shoots in the ground (provignage in French, Vergruben in German), soon led to an irregular appearance of the plantation layout even when the vines were originally planted in rows (*vigne en foule* in French). This layout precluded the use of animal-driven ploughs in the vineyards[1], so the soil had to be hoed several times a year. The high labour required to cultivate this kind of vineyard meant that a single person could not properly work more than one hectare of vines over the year, a whole family no more than two hectares.

This way of cultivating vines contrasts completely with the Mediterranean practice of intercropping, the most pronounced form of which is the coltura promiscua of the central and north-eastern Italian regions. Here vines were planted together with trees in the middle or on the fringes of arable fields. The number of vines per hectare varied between 100 and 1500. They were pruned only lightly, if at all, and needed no special support since they climbed on the trees. The soil could be worked with ploughs. Over the year, a single plot yielded cereals, vegetables, grapes, and tree crops, or fire wood and animal fodder, depending on the nature of the trees planted. This form of wine-growing dominated Italian viticulture until the middle of the twentieth century though it has been disappearing since then. In 1957, after half a century of regression, there were still 2,723,000 ha of *coltura promiscua* (corresponding to 20 percent of land used for cereal production) and 1,088,000 ha of monocultural vineyards. (Desplanques 1959). It can still be encountered in northern Portugal.

Variants of this extensive way of cultivating vines existed under different names all over the Mediterranean area and at its fringes: *alberata* (Tuscany, Umbria), *piantata* (Emilia Romagna, Veneto), *arbustata* (Campagna), *vignaadalteno* (Piemont); *ouillière* (Provence), *joualle* (southwestern France), *vigne en hautins* (Dauphiné), *hutins* (in the region of Geneva, which seems to mark the most northern extension of this form of wine-growing), and the like (Desplanques, 1959; Rapetti, 1984: 16-17; Feniello, 2005: 74-75; Lachiver, 1988: 186-187; Zumkeller, 1992: 110, 186-187; Vilar, 1962: I, 297-299). But even where monocultural vineyards predominated in the Mediterranean area, the density of the plants was much lower and the soil was often worked with ploughs, as in Sicily (2,000-4,000 vines per hectare) or in Castille (1000-4000 vines per hectare), and yields were therefore much lower than in more northern regions.(Bresc, 1986:180; Huetz de Lemps, 1991:152, 162; Antoine, Boehler, Brumont, 2000: 217).

1 The use of the plough and planting in rows began in most northern wine-growing regions only with the reconstitution of vineyards after the phylloxera epidemic, which made provignage impossible. Reconstitution after phylloxera might also have occasioned a first shift from intercropping to monocultural vineyards in the areas of *coltura promiscua* (Galassi, 1986: 106-110).

It is puzzling to find that in most of these regions, specialised (monocultural) vineyards existed alongside polycultural fields. In the case of Italy this is complicated by the peculiar geographic distribution of the *coltura promiscua* system. Emilio Sereni has proposed a genetic explanation of this pattern, associating the regions of monocultural vineyards and low trimmed vines with the areas of Greek colonisation, and the *coltura promiscua* regions with the area of Etruscan cultural influence (Sereni, 1981). Others have given agro-climatic explanations: on the plains, vines are associated with trees; on the hill slopes, they are found in specialised vineyards. Is there also a functional logic in the location and spatial distribution of the two ways of Mediterranean wine-growing? Roman civilisation, had had both forms under the names of *vinea* and *arbustum gallicum* (the attribute *gallicum* refers to Gallia Cisalpina), but after the decline of Roman civilisation, the link between vines and trees grew weaker during the early Middle Ages. A new phase of expansion began in the later Middle Ages under the influence of the city-states of central Italy which promoted the planting of trees and vines on their territories. In late medieval Tuscany dispersed settlement, the development of compact sharecropping holdings (*mezzadria poderale*), and the extension of the *coltura promiscua* system appear to be closely connected (Pinto, 1982; Biagioli, 2010). However, the triumph of the *coltura promiscua* did not completely end specialised wine-growing. Indeed, it survived or expanded near consumption centres such as Florence (De la Roncière, 1978) or Mantua (Sereni, 1997: 97). The same development seems to have taken place in the region around Naples, driven by growing wine exports during the fifteenth and sixteenth centuries (Feniello, 2005: 186-196). In a contrary development however the expansion of wine-growing in Piedmont during the late Middle Ages brought about a shift from monocultural vineyards (*vigna a spanna*) to intercropping (*alteno*) (Rapetti, 1984: 16-17; Donna D'Oldenico, 1975). Hence, the mode and intensity of cultivation varied with market conditions. As the extensive form of wine-growing, *coltura promiscua*, progressed in the late medieval and early modern period, forms of cultivation intensive in labour and other costs continued only where demand for wine was great.

Because extensive wine production had become widespread, wine was cheap and readily available for daily consumption in the Mediterranean area. Table 13.2 displays a sample of relative prices of wine (1 hl of wine/1 hl of wheat or rye) in major towns of the Mediterranean and extra-Mediterranean areas. As far as possible, only prices for the locally-produced or the lowest-priced wine have been taken into consideration. It is evident that in the Mediterranean area wine was cheaper than an equivalent measure of bread-grains whereas its relative price rose the further north one moved[2].

[2] Paris seems to be the exception to the rule but this could be explained by its ample supply of wine from growing regions in the vicinity and the high level of cereal prices.

It became very dear in regions where it had to be imported over large distances. Wine in the Mediterranean area was a staple good that was firmly integrated into the daily diet of the common people whereas in the north it was a kind of luxury, at least outside the growing regions. Here it was consumed occasionally but then, when available, often also to excess (Braudel & Spooner, 1967: 410).

Table 13.2. Relative prices of wine in major European towns

	Aix-Provence (1)	*Béziers (2)*	*Paris (3)*	*Florence (4)*	*Milan (5)*	*Vienna (6)*	*Munich (7)*	*Frank-furt (8)*	*Lier (Brabant) (9)*
				(1 hl wine/1 hl of wheat or rye)					
1550/59				0.69					
1560/69			0.99	1.22				2.71	
1570/79	0.62		0.99			1.53	5.35		7.23
1580/89	0.63		1.24			2.92			
1590/99	0.59	0.61	1.29			2.97			8.75
1600/09	0.58					2.29			
1610/19	0.88			1.04		2.67	10.24		
1620/29	0.40								
1630/39	0.72		1.11			1.77			
1640/49	0.60		1.28			1.58		4.93	
1650/59	0.46		1.21			2.56			
1660/69	0.38		1.05			1.61			
1670/79	0.42		1.27						
1680/89			1.09			2.01			
1690/99			1.19			2.42			
1700/09					1.03	1.78			
1710/19					0.85	2.02			

Sources. (1) Baehrel (1961: 558-561: wine/rye). (2) Le Roy Ladurie (1966: 820-823: wine/wheat). (3) Baulant (1976: 961: wine/wheat: 1561-1565, 1571-1575, 1585-1589, 1594-1598); Baulant (1968: 539: wheat 1630/1639-1690/1699). Lachiver (1982: 861-862: vin d'Argenteuil 1630/1639-1690/1699). (4) Parenti (1939: App. 7, 11: wine/wheat). (5) De Maddalena (1974: 428, 435: vino crudello/wheat 1701-1710, 1711-1720). (6) Pribram (1938: I, 570, 575-576: low quality wine = wine purchased/rye). (7) Elsass, (1936: I, 542-543: Neckar wine/rye). (8) Elsass (1936: II, 494-495: Rhine wine/rye). (9) Van der Wee (1963: I, 187-188, 299: Rhine wine/wheat).

III. Specialisation and markets in extra-Mediterranean viticulture

If you cannot sell your wine, why have a large vineyard?[3] Present-day explanations of the link between location and wine quality glorify the terroir. In contrast to this cultural bent, Roger Dion appositely recalls this line from the Olivier de Serres' Théâtre d'agriculture (1600), using the French case as the basis for a radically different interpretation of extra-Mediterranean viticulture. Dion argues that, since well-exposed hillsides, well-drained and warm soils exist all over the area of extra-Mediterranean viticulture, these physical conditions constituted no more than a potential which, in the past, could only be realized by the combination of transport facilities and commercial organization (Dion, 1959). A short glimpse at any map indicating the geographical distribution of extra-Mediterranean wine-growing regions confirms Dion's interpretation. The river valleys of the Garonne and Dordogne, of the Loire, Seine, Marne, Yonne and Rhône in France, of the Rhine, Main, Neckar and Moselle in Germany and of the Danube in Austria – to name just the most important – were, and are often still, packed with vineyards. Even the Elbe in Bohemia and the Saale and Unstrutt in Thuringia determined the situation of the rare wine-growing areas found in these parts of Europe. Orléans and Blois, Paris and Reims, Cologne, Strasbourg, Frankfurt and Vienna were major trading places for wine. The quantity of wine exported through the ports of Bordeaux, Nantes and La Rochelle amounted to hundreds of thousands of hectolitres in the late medieval and early modern period (Le Roy Ladurie, 1993: 779-781).

For the sake of brevity I will limit myself to a closer look at two examples. During the late medieval and early modern period, the Danube and its tributaries were the main transport routes for wine exported from lower Austria to Upper Austria, Bavaria, Salzburg and Tyrol. Boats transporting wine had to be drawn upwards against the river current, and all along its course boatmen and shipping guilds were kept busy transporting wine. Further to the southeast of Vienna the rich wine-growing lands of Western Hungary discouraged shipments in the opposite direction. Meanwhile the Estates of Lower Austria blocked all attempts to sell Hungarian wine on the markets upstream until 1756, allowing only the overland transit of limited quantities to Bohemia and Silesia. Before the Thirty Year's War about 100,000 hl of wine a year were exported from Vienna on the Danube and, as the river reached the border between Lower and Upper Austria, produce of the wine-growing regions along the river swelled total shipments to 200,000 hl. About half of it was consumed in Upper Austria or was transported overland from Linz to Salzburg; the other half went on to Bavaria (80,000 hl per year in the second half of the sixteenth century) (Landsteiner, 1996: 28). The amount of wine shipped on the Danube is therefore not

[3] 'Si (vous) n'êtes pas en lieu pour vendre votre vin, que feriez-vous d'un grand vignoble?'

much different from the quantity leaving Bordeaux (200,000-250,000 hl) or Nantes (150,000-180,000 hl) by sea in the sixteenth century (Le Roy Ladurie, 1993: 779-781; Lachiver, 1988: 123, 134).

In 1705, French government officials, seeking to levy duties, compiled lists of brokers acting as middlemen between buyers and sellers of wine, or buying up the product on commission for wine merchants. Plotting this data on a map, Thomas Brennan was able to depict the principal arteries of the interregional wine trade (Brennan, 1997: 94-100). The main trading and export centres, with Bordeaux on top followed by Reims and Nantes, stand out in this map, and the role of rivers and ports for the interregional and international trade in wine is clearly visible. The city of Beaune in Upper Burgundy is the only major wine trading centre on Brennan's map not located on a river or in its vicinity. Wine-growing regions having no direct access to water transport were heavily disadvantaged under pre- industrial transport conditions. The only way to compensate for that was to raise the quality standards for wine produced in such regions. Since the economic feasibility of transport over longer distances depended on the relation between the cost of transport and the value of the cargo (see Landers, 2003: 86-90) only high priced wine could support the costs of overland transport. Wines from the Côte d'Or rose to the top of the hierarchy of French wines during the late medieval period because of measures taken by noble and bourgeois vineyard owners, with the support of the dukes of Burgundy: in 1395 the planting of gamay grapes was banned by an edict. The union of Burgundy with Flanders, one of the principal export markets under the Valois, guaranteed the commercial success of the 'wines of Beaune' despite the adverse transport situation (Dion, 1959: 285-300; Craeybeckx, 1958: 45-77).

Although the fame of its wines paled by comparison with those from Beaune, the county of Retz on the border between Lower Austria and Moravia was in a quite similar position. The town dominated an area comprising nine villages, whose territories were thickly planted with vines. To protect the reputation of the local *cru*, no foreign wine was allowed to enter the district. Only burgesses of the town bourgeoisie, who themselves owned an important part of the vineyards, had the right to store their own harvest, and the wines they bought in the nearby villages, in the extensive network of cellars under the town. Every year in autumn and in spring long caravans of horse-drawn wagons arrived from the towns of Bohemia and Moravia to carry the barrels, marked with the the town of Retz' own coat of arms, to the places of consumption (Landsteiner, 1992: I, 84-90; Landsteiner, 1996: 40-42).

One further force determining the market for wine and the geography of viticulture, with major consequences for the kind of wine produced, was the growth of capital cities during the early modern period. Vienna and Paris are good examples of this development. Medieval viticulture in the extra-Mediterranean macro-region was to a very high degree dominated by towns and their inhabitants (Dion, 1959: 41). Vienna was a wine-producing and wine-exporting town from its origins down to the middle of the seventeenth century, and winegrowing was by far the most important sector of the urban economy. At the end of the sixteenth century a third of all inhabitants of the town and its suburbs were liable to a tax on owners of vineyards. The winegrowing district of Vienna was the most important and most prestigious wine-growing area in Lower Austria. The right to bring wine into the town and to sell it was limited to the strictly defined boundaries of an area that coincided with the region where bourgeois property predominated (Landsteiner, 2003: 141-146). Wine from Vienna, especially that destined for export on the Danube, was highly priced and must therefore have been of a certain quality. But everything changed when the town population grew as it became the capital of a growing empire in Central Europe. Population numbers more than trebled in the course of the seventeenth century from 30,000 around 1600 to 110,000 in 1710 and attained 220,000 at the end of the eighteenth century (Weigl, 2003: 110). Under the pressure of demand for cheap wine by this growing population the old system of wine-market regulation broke down. In the course of the seventeenth century, Vienna was transformed from a wine-producing into a wine-consuming town. According to the records of the tax on wine consumption, at the end of the sixteenth century only 45,000 hl were sold retail, whereas in 1730-1731 435,000 hl of wine from Lower Austria and 7,100 hl of 'foreign' wine paid the entry tax (Landsteiner 2003: 145). Consequently the entire agricultural landscape of northwestern Lower Austria (the so-called Weinviertel – the "quarter of wine") was transformed. Peasants all over this area began to plant a new, high-yield grape vine and started to dig cellars outside their villages, which today are a picturesque element of the agrarian landscape (Landsteiner, 2002). Meanwhile the once-prestigious suburban vineyards were rooted up as the growing capital of the Hapsburg monarchy pushed out its boundaries.

Those familiar with the history and historiography of French wine-growing will easily recognize the parallels between this story and the effect of growing demand for wine in Paris on wine-growing in the Île-de-France and indeed mass production of cheap wines in the upper Loire valley and along the Yonne. The measures taken by the Parlement of Paris to secure the city's wine supply included an ordinance, promulgated in 1577, which forbade wine merchants and tavern-keepers from buying wine for the Parisian market within a radius of 88 km around the city. This policy was supposed to protect the vineyards near the city from the harmful effects of the mass market, but it was ineffective in halting the switch to mass production in the region (Dion,1959: 497-576; Lachiver, 1988: 314-324).

IV. Specialisation and the social relations of wine production

There was no uniformity in the relations of production prevalent in extra-Mediterranean viticulture. Vineyards worked by their owners, by sharecropping and by wage labour existed side by side and the proportions varied from region to region (Carmona & Simpson 2010: Laurent, 1958: I, 259). Nevertheless, some tendencies are clearly discernible. Forced labour was rare even in regions such as east-central Europe, where manorial structures continued to dominate rural society during the early modern period and forced labour was common in the cultivation of seigneurial demesnes. Because wine-growing required diligent and motivated workers, it would have been very risky to use coerced labour for all but simple tasks such as grape picking. Lords therefore refrained from using corvée labour and preferred to pay workers a wage for their services, even if it was often below the market rate, or to let them cultivate vines themselves under sharecropping. Ecclesiastical and lay landlords clearly preferred collecting rents in wine to producing wine themselves on their domains (Landsteiner 2011: 288-291).

Cultivation by wage labour was quite frequent in suburban vineyards where bourgeois property prevailed. But this entailed high supervision costs and required a quality product. High labour intensity drove up costs and unstable yields in northern regions could easily lead to financial loss when the harvest was poor. Vines, especially in regions devoted to quality production, represented a considerable capital, and they required continual effort by the vigneron. Short-term leases were therefore avoided and long-term labour contracts, such as *prix-faitage*, where a qualified worker tended a vineyard over the course of the working year in return for a fixed wage, were common in regions producing quality wines for wider markets. These regions were typically dominated by towns and urban bourgeoisies (e.g. the Bordeaux region, Paris, Champagne, Burgundy, Alsace and the Upper Rhine region in general, the Vienna region and the like (Pijassou, 1980: I, 343-346; Lachiver, 1982: 41-42; Laurent 1958, I: 259-263; Schulz 1985: 358-361, 436-437; Landsteiner 1992: I, 64-69, 200-210). The towns developed institutions to supervise the workers and conflicts between labourers and vineyard owners over working hours and wages were frequent (Delafosse, 1948; Landsteiner, 1993).

Vines are very sensitive to poor cultivation and a negligent vigneron put in danger the capital embodied in the plants. As a result there were virtually no fixed money rent contracts, in winegrowing, least of all short-term ones. In regions where quality production was the rule, vineyard owners also generally took an interest in the product itself. Landowners unable or unwilling to manage cultivation themselves preferred sharecropping contracts to keep down supervision costs and share the risk of unstable yields with their tenants. This way of organizing the cultivation of vineyards was common all over the Mediterranean area, in south-western France, in Burgundy south of Beaune, the Beaujolais and the region of Lyon, along the middle and lower Rhine region and the Moselle, and in a stretch of the Danube valley in Austria where monastic property prevailed. It was applied either to the cultivation of single vineyards or to entire wine-growing estates (Laurent, 1958: I, 260-263; Garrier, 1973: I, 149-152; Durand, 1979: 248-250; Goujon, 1976; Volk, 1993:92-97; Landsteiner, 2001). Sharecropping is usually thought to keep down transaction costs associated with monitoring, and negligence, and to reduce the risk of specialisation in cash crops. Yet these cost advantages are counterbalanced by costs associated with the division of the crop, although this problem could be got around by locating grape processing in the landowner's vinery (Carmona & Simpson, 2010). Sharecropping was frequent in regions with a low degree of specialisation where multi-cropping or intercropping prevailed and where quality was low.

Although it is impossible to provide comprehensive quantitative proof it seems that the most common form of wine-growing was direct cultivation by small landowners (Laurent, 1958: II, 148-168; Gavinaud, 1983; Lachiver, 1982; Landsteiner & Langthaler, 1993; Rapetti, 1984: 65-96). It has been often observed that wine-growing leads to the spread of landed property among peasants and the common people. Planting contracts such as the French *complant* (Dion 1959: 202-205) or the Catalan *rabassa morta* (Vilar, 1953) made vineyard property accessible to the lower classes. Moreover, it was simply in the interest of landlords to hand over swathes of land otherwise useless for cultivation to peasants, who could plant them with vines, in return for favourable property rights and rents in kind, such as the *Bergrecht* (implying hereditary tenancy of vineyard plots) in Lower Austria and Styria (Landsteiner 1992: I, 25-28). In the right circumstances viticulture therefore promoted a 'smallholder society' (Netting 1993). Easy access to land, high labour intensity and minimal capital investments for cultivation, partible inheritance practices, a very lively land market and a low degree of social differentiation all characterized wine-growing societies in many regions (Labrousse, 1944: 554-559, 594-600; Goubert, 1986: 129- 132; Brennan, 1997:8-25; Carmona & Simpson, 2010).

Low entry costs and the reliance of smallholders on their family labour force made supervision easier, and reduced moral hazard. These factors seem to have determined the dominance of small owner-operators in many winegrowing regions. How did these producers cope with the risks of cash-crop production? The distribution of landed property in two Lower Austrian wine-growing villages at the end of the eighteenth century provides a first answer (Table 13.3).

Table 13.3. Land ownership pattern in two Lower Austrian wine-growing villages (Ober- and Mitterretzbach), 1785

	Average size of holding per decile				
deciles	*total land (ha)*	*vineyards (ha)*	*arable (ha)*	*grassland (ha)*	*estimated wine yield (hl)*
1.	0,15	0,15	0,00	0,00	2,2
2.	0,23	0,22	0,01	0,00	2,8
3.	0,37	0,30	0,03	0,04	3,8
4.	0,55	0,50	0,06	0,00	6,6
5.	0,81	0,75	0,06	0,00	9,8
6.	1,23	1,00	0,18	0,05	14,2
7.	1,92	1,48	0,37	0,07	21,1
8.	2,94	1,68	1,06	0,20	24,4
9.	4,70	2,60	1,83	0,27	39,9
10.	8,75	3,90	4,29	0,56	62,6
(n=218)					
average size	2,12	1,24	0,77	0,12	18,5
Gini-coeff.	0,577	0,498	0,758	0,770	0,537

Source. LANDSTEINER & LANGTHALER (1997: 211).

Close to the town of Retz, mentioned above, these two villages of Ober- and Mitterretzbach were closely so fused together in a densely populated region that we can treat them as one. They numbered 218 landowners in 1785, when the cadaster containing the data in Table 13.3 was drawn up. Only 107 could owned a house; the rest lived as lodgers in the other peoples' houses; but virtually every adult inhabitant owned at least a small parcel of land planted in vines. The territory of the two villages contained 270 ha of vineyards, 168 ha of arable land and 25 ha of grassland. Animal husbandry was minimal. House-owners also usually had one or two cows and a few pigs. Draught animals were extremely rare and so were ploughs. Smallholders worked their small parcels of arable land with hoes just like their vineyards or exchanged their labour with their wealthier neighbours in return for help in ploughing. Vineyard property was clearly more evenly distributed among the villagers than arable land or the few meadows.

Those located in the lower half of the distribution pattern were mainly lodgers and owned less than a hectare of land completely planted in most cases with vines. By modern standards, probably all holdings, with the possible exceptions of those at the top of the distribution, were 'specialised' wine-producing holdings. But in the lower half of the distribution this specialisation had grown out of poverty. These micro-properties could in no way provide an income to sustain a family. Their owners had to combine it with wage-labour for the wealthier landowners in the village. As in most other intensive wine-growing regions there was no craft production or proto-industrial activity on a larger scale (Mendels, 1980). Only the few who could combine a cash-crop (wine) with food-crop production (cereals) can have lived comfortably in these economic circumstances (Landsteiner & Langthaler, 1997, and Landsteiner, 1999a, for more detailed analyses).

Benoît Musset's detailed analysis of mid-eighteenth century landholding patterns in those parts of the Champagne region where quality wine was produced and wine-growing was the main activity shows a structure very similar in specialisation, size of holdings and distribution of landed property. Owners with less than one hectare of land relied completely on wine-growing whereas larger holdings appear much more diversified (Musset, 2008: 399-414).

These examples suggest that small-scale producers who had the necessary resources avoided full specialisation and, consequently, complete market dependence. In extra-Mediterranean wine-growing areas it was only in particular circumstances that specialisation at the level of the holding dominated. Market structure and market access seem to have been more decisive than relations of property and production (Brennan, 1997, has convincingly brought this aspect into the discussion). An indispensable condition for specialisation and quality production was access to markets with buyers able and willing to pay for quality (Enjalbert, 1953). In this respect the wealthy producer not entirely dependent on the quick sale of his harvest was certainly in a better position than the small-scale and highly specialised producer who was forced to sell at any price to the best buyer available. There was a further important distinction: those who produced the grapes were not necessarily the same people who made the wine. Producing the raw material, processing, and finishing the final product were more often than not separate steps of a commodity chain in which institutionalized and informal monopolies played a major role.

V. The risks of specialisation in a pre-industrial environment

Several elements arising from the climatic vulnerability of extra-Mediterranean viticulture and the basic structures of the *Ancien Régime* economy' made specialized wine-growing risky for producers. First of all, wine yields fluctuated widely due to unstable weather conditions. Figure 13.1 gives an idea of these oscillations; it depicts the deviation from trend of yields per hectare in what was probably the largest wine-growing enterprise in early modern Lower Austria. This was the Vienna City Hospital, which operated between 70 and 100 ha of vineyards in this period (Landsteiner 1992: I; 194-196).

Figure 13.1. Wine yields per hectare of the Vienna City Hospital, 1550-1680

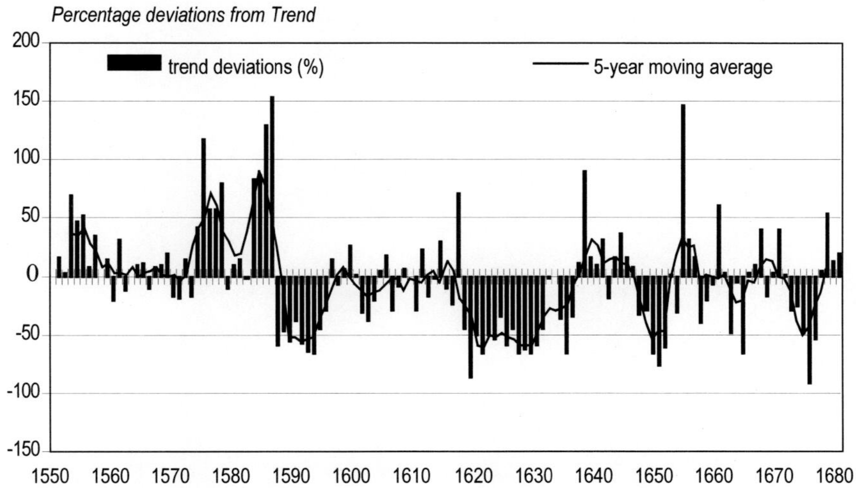

Source. Archives of the City of Vienna (VCA), Hospital accounts, 1550-1680.

The unstable volume of grape harvests was the single most important factor in determining the net revenues of this wine-growing estate. In addition climate change in the period under review made this insecurity even worse. The cooling phase of the central European climate in the last third of the sixteenth century caused a long series of harvest failures, beginning in 1587. Yields per hectare suddenly dropped from 45 hl/ha in 1586 – an extremely rich harvest – to 7 hl/ha in 1587 and remained at this low level for seven years. Wine-growing all over Central Europe was hit badly by this shock (Landsteiner 1999b)[4].

Rising prices for wine could not make up revenue loss caused by low yields, since price elasticity of demand for wine was high in this part of Europe and substitutes such as beer were ready to hand. Beer sales soared to unprecedented levels in Vienna at the end of the sixteenth century, and all over the countryside noble landlords took to brewing more beer. The switch from wine to beer in the consumption habits of

4 The low yields of the 1620s and 1630s, which also stand out in the diagram, were partly caused by the ravages of the Thirty Year's War.

the popular classes in this part of Europe might well have begun at this time. The Vienna City Hospital used wage labourers to cultivate its vineyards. As their wages amounted to 70 per cent of the variable costs of production, the institution ran heavy deficits in those years (Figure 13.2).

Figure 13.2. Net revenues of wine production, wine yields per hectare, and wine prices in Vienna, 1550-1600

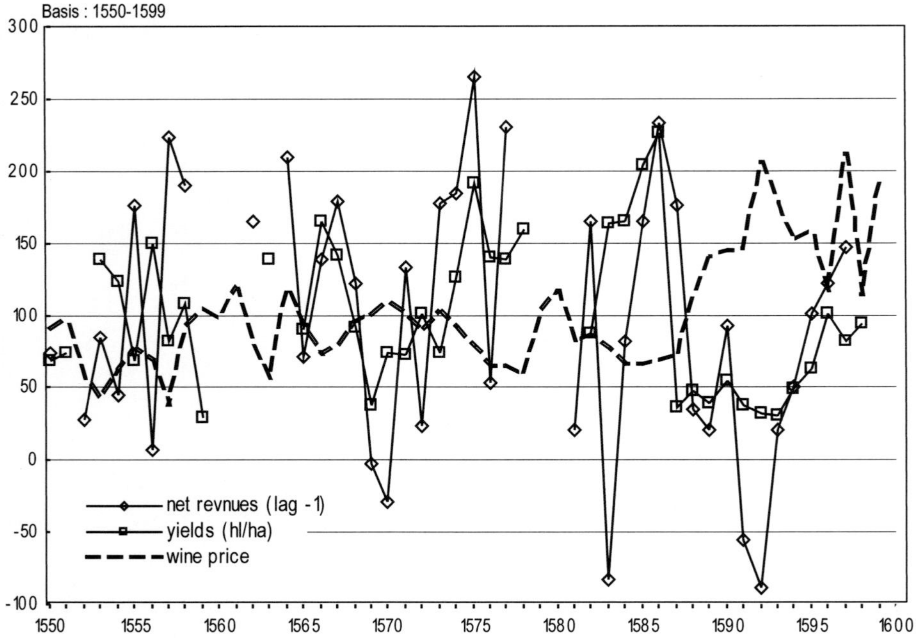

Source. VCA, Hospital accounts, 1550-1600.

By drawing on its stock of wine a big enterprise like the hospital could make up somewhat for its losses on poor harvests. Smaller producers, who had to sell their harvest shortly after bringing it in, could not. For them such a succession of harvest failures was a disaster.

The hospital's deficits of 1570 and 1571, in years when yields were not particularly low, points to another factor that made the market for wine so unstable. The years around 1570 were a period of very poor grain harvests, which triggered a severe subsistence crisis in large parts of Central Europe (Abel, 1974: 70-98). Figure 13.3 points up the consequences of this crisis for the wine market.

Figure 13.3. Net returns of wine production, wine yields (hl/ha), wine and rye prices

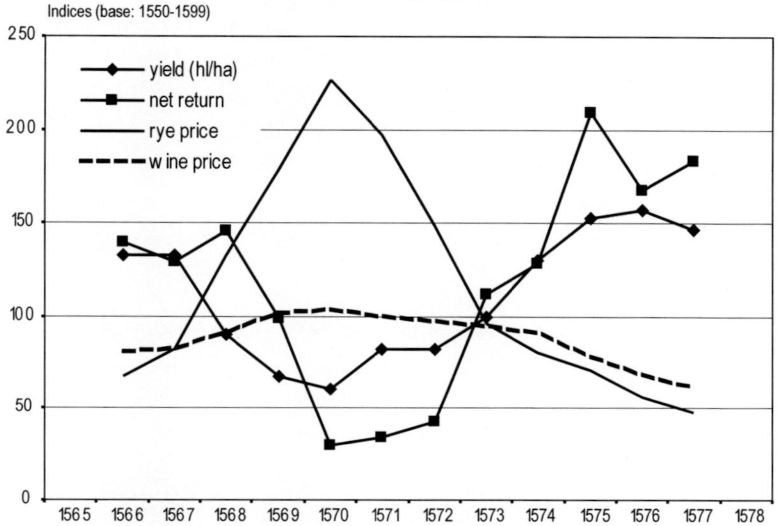

Source. VCA, Hospital accounts, 1565-1578.

The diagram shows a marked coincidence between steeply rising rye prices and the hospital's declining net income from wine production. Wine yields were low as well in this period, but the price of wine responded only a little to these conditions.. In an early modern economic context, where a large swathe of the population spent much of their income on basic subsistence goods, demand price elasticity for bread was very low. It is not unreasonable to suppose that the demand by the poorest groups of the population rose in such circumstances since they had to spend all their income on this basic good (Simonin, 1980; Grenier, 1996: 307-315). Deficient harvests of bread grains therefore drew purchasing power away from all other goods to cereals. A kind of 'Labrousse effect' affected the market for wine and demand, especially demand for cheap wines, slumped (Labrousse, 1944: 296-298). It has been stressed at the beginning of this paper that wine was a kind of everyday luxury in extra-Mediterranean Europe and not, as in the Mediterranean area, a fixed part of the daily diet. The market for quality wines was much less hit by these mechanisms, since income elasticity of demand of consumers there was much lower, and those groups able to offer cereals on the market (large producers and landlords receiving rents in kind) might even have profited from the situation by a transfer of purchasing power (Simonin, 1992).

VI. Conclusion

Small-scale producers of cheap wines for mass consumption, who drew most of their income from the sale of their wine harvest, were doubly hit by the situation: as sellers of wine and as buyers of cereals or bread. Being completely dependent on income from the sale of grapes or wine – in other words, being specialized – in such circumstances was a dangerous way of life. No wonder that those who could avoid it, did so.

Bibliography

ABEL, William (1974), *Massenarmut und Hungerkrisen im vorindustriellen Europa*, Hamburg and Berlin, Parey, 427 p.

ANTOINE, Annie, BOEHLER, Jean-Michel, BRUMONT, Francis (eds) (2000), *L'Agriculture en Europe à l'époque moderne*, Paris, Belin, 447 p.

BAULANT, Micheline (1968), 'Le Prix des grains à Paris de 1431 à 1788', *Annales Economies Sociétés Civilisations*, p. 520-540.

BAULANT, Micheline (1976), 'Prix et salaires à Paris au XVIe siècle', *Annales Economies Sociétés Civilisations*, p. 954-995.

BAEHREL, René (1961), *Une Croissance. La Basse-Provence de la fin du XVIe siècle à 1789*, Paris, SEVPEN, 842 p.

BIAGIOLI, Giuliana (2010), 'The Evolution of Property Rights in Tuscany from the End of the Middle Ages to the Nineteenth Century', in Rosa CONGOST, Rui SANTOS (eds), *Contexts of Property in Europe. The Social Embeddedness of Property Rights in Land in Historical Perspective*, Turnhout, Brepols, Rural History in Europe, 5, p. 61-82.

BRAUDEL, Fernand (1966), *La Méditerranée et le monde méditerranéen à l'époque de Philippe II*, Paris, A. Colin, 589 p.

BRAUDEL, Fernand, SPOONER, Frank (1967), 'Prices in Europe from 1450 to 1750', in Edwin Ernest RICH, Charles H. WILSON (eds), *The Cambridge Economic History of Europe*, vol. IV, *The Economy of Expanding Europe in the Sixteenth and Seventeenth Centuries*, Cambridge, Cambridge University Press, p. 376-486.

BRENNAN, Thomas Edward (1997), *Burgundy to Champagne. The Wine Trade in Early Modern France*, Baltimore-London, J. Hopkins University Press, XXI-350 p.

BRESC, Henri (1986), *Un Monde méditerranéen. Économie et société en Sicile, 1300-1450*, Rome, École française de Rome, Paris, diffusion de Boccard - Palermo, Accademia di scienze, lettere e arti di Palermo, 2 vol. 981 p.

CARMONA, Juan & SIMPSON, James (2010), 'Explaining Contract Choice: Vertical Coordination, Sharecropping, and Wine, France 1850-1950', Universidad Carlos III de

Madrid, Working Papers in Economic History, WP 10-15 (http://www.uc3m.es/uc3m/dpto/HISEC/working_papers/_general.html)

CRAEYBECKX, Jan (1958), *Un Grand commerce d'importation: les vins de France aux anciens Pays-Bas (XIIIe-XVIe siécle)*, Paris, SEVPEN, XXXII-314 p.

DE LA RONCIÈRE, Charles M. (1978), 'Le Vignoble florentin et ses transformations au XIVe siècle', in *Le vin au Moyen Âge : productions et producteurs,* (Actes du IIe Congès des Médiévistes, Grenoble, 1971), Grenoble, Presses Universitaires de Grenoble, p. 125-154.

DE MADDALENA, Aldo (1974), *Prezzi e mercedi a Milano dal 1701 al 1860*, Milano, Banca commerciale, 455 p.

DE VRIES, Jan (1994), 'The Industrial Revolution and the Industrious Revolution', *Journal of Economic History 54*, p. 249-270.

DELAFOSSE, Marcel (1948), 'Note d'histoire sociale. Les vignerons d'Auxerrois (XIVe-XVIe siècle)', *Annales de Bourgogne,* 20, p. 7-41.

DEMONET, Michel (1990), *Tableau de l'agriculture française. L'enquête de 1852*, Paris, École des Hautes Études en Sciences Sociales, 304 p.

DESPLANQUES, Henri (1959), 'Il paessagio rurale della coltura promiscua in Italia', *Rivista Geografica Italiana,* 66, p. 29-64.

DION, Roger (1959), *Histoire de la vigne et du vin en France des origines au XIXe siècle*, Paris, Flammarion, XII-771 p.

DONNA D'OLDENICO, Giorgio (1975), 'Vitibus spannis et altinis. In margine a la civiltà della spanna da Gattinara a Lesiona, *Rivista di Storia dell'agricoltura,* 15, p. 75-80.

DURAND, Georges (1979), *Vin, vigne et vignerons en Lyonnais et Beaujolais (XVIe-XVIIIe siècles)*, Lyon-Paris, Presses Universitaires de Lyon, École des Hautes Études en Sciences Sociales, 540 p.

ELSASS, Moritz John (1936), *Umriss einer Geschichte der Preise und Löhne in Deutschland*, Leiden, A.W. Sijthoff, 3 vol. X-808, 649, VIII-138 p.

ENJALBERT, Henri (1953) 'Comment naissent les grands crus. Bordeaux, Porto, Cognac', *Annales Économies Sociétés Civilisations*, p. 315-328, 457-474.

European Commission, Directorate-General for Agriculture and Rural Development (2006), 'Wine–Economy of the Sector'. http://ec.europa.eu/agriculture/markets/wine/studies/rep_econ2006_en.pdf.

FENIELLO, Amadeo (2005), *Les Campagnes napolitaines à la fin du Moyen Âge*, Rome, École française de Rome, IX-287 p.

GALASSI, Francesco L. (1986), 'Reassessing Mediterranean Agriculture: Retardation and Growth in Tuscany, 1870-1914', *Rivista di Storia Economica,* 3, p. 91-121.

GARRIER, Gilbert (1973), *Paysans de Beaujolais et du Lyonnais, 1800-1970*, Grenoble, Presses Universitaires de Grenoble, 2 vol., 714 et 246 p.

GAVIGNAUD, Geneviève (1983), *Propriétaires-viticulteurs en Roussillon. Structures, conjunctures, société, XVIIIe-XXe siècle*, Paris, Publications de la Sorbonne, 2 vol., 1-788 p.

GOUBERT, Pierre (1986), *The French Peasantry in the Seventeenth Century*, Cambridge, Cambridge University Press, XI, 242 p.

GOUJON, Pierre (1976), 'Contribution à l'étude de la grande propriété dans le vignoble mâconnais au XIXe siècle, *Revue d'Histoire Économique et Sociale,* 54, p. 371-405.

GRANTHAM, George (1999), 'Contra Ricardo: On the Macroeconomics of Pre-Industrial Economics', *European Review of Economic History,* 3, p. 199-232.

GRENIER, Jean-Yves (1996), *L'Économie d'Ancien Régime, un monde de l'échange et de l'incertitude,* Paris, Albin Michel, 489 p.

HORDEN, Peregrine, & PURCELL, Nicholas (2000), *The Corrupting Sea. A Study of Mediterranean History*, Oxford, Blackwell, XIII-761 p.

HUETZ DE LEMPS, Alain (1991), 'Vignobles et vins de Castille du XVIe au XVIIIe siècle', in *Le vigneron, la viticulture et la vinification en Europe occidentale au Moyen Âge et à l'époque moderne* (Centre Culturel de L'Abbaye de Flaran, XIe Journées internationales d'histoire), Auch, p. 147-188.

LABROUSSE, Ernest (1944), *La Crise de l'économie française à la fin de l'Ancien Régime et au début de la Révolution*, Paris, Presses Universitaires de France, LXXV-664 p.

LANDERS, John (2003), *The Field and the Forge. Population, Production, and Power in the Pre-Industrial West*, Oxford, Oxford University Press, XIII-440 p.

LANDSTEINER, Erich (1992), 'Weinbau und Gesellschaft in Ostmitteleuropa. Materielle Kultur, Wirtschaft und Gesellschaft im Weinbau, dargestellt am Beispiel Niederösterreichs in der frühen Neuzeit', unpublished doctoral thesis, University of Vienna.

LANDSTEINER, Erich (1993), 'Einen Bären anbinden', *Österreichische Zeitschrift für Geschichtswissenschaften,* 4/2, p. 218-252.

LANDSTEINER, Erich (1996), 'Weinbau und bürgerliche Hantierung. Weinbau und Weinhandel in den landesfürstlichen Städten und Märkten Niederösterreichs in der frühen Neuzeit', in Ferdinand OPLL (ed.), *Stadt und Wein* (Beiträge zur Geschichte der Städte Mitteleuropas 14), Linz, p. 17-50.

LANDSTEINER, Erich (1999a), 'Household, Family and Economy among Wine-Growing Peasants. The Case of Lower Austria in the First Half of the 19th Century', *The History of the Family,* 4, p. 113-135.

LANDSTEINER, Erich (1999b), 'The Crisis of Wine Production in Late Sixteenth-Century Central Europa: Climatic Causes and Economic Consequences', *Climatic Change,* 43, p. 323-334.

LANDSTEINER, Erich (2001), 'Teilbau und Lohnbau. Zwei Formen der Arbeitsorganisation im Weinbau des Kremser Raumes', in Willibald ROSNER (ed.), *1000 Jahre Krems – am Fluß der Zeit*, St. Pölten, Studien und Forschungen aus dem Niederösterreichischen Institut für Landeskunde, 13, p. 186-220.

LANDSTEINER, Erich (2002), 'Cellars in Town and Countryside. Wine Storage Architecture and the Social Relations of Production in East Central Europe', in *Douro – Estudos & Documentos,* 13, p. 125-133.

LANDSTEINER, Erich (2003), 'Wien – eine Weinbaustadt?', in Peter CSENDES, Ferdinand OPLL (eds), *Wien. Geschichte einer Stadt*, Wien, Böhlau, vol. II, p. 141-146.

LANDSTEINER, Erich (2011), 'Demesne Lordship and the Early Modern State in Central Europe: the Struggle for Labour Rent in Lower Austria in the Second Half of the Sixteenth Century', *Agricultural History Review,* 59/2, p. 266-292.

LANDSTEINER, Erich, LANGTHALER, Ernst (1997), 'Ökotypus Weinbau: Taglöhner-oder, Smallholder-Gesellschaft', in Franz X. EDER, Peter FELDBAUER, Erich LANDSTEINER (eds), *Wiener Wege der Sozialgeschichte*, Wien, Böhlau, Institut für Wirtschafts und Sozialgeschichte, p. 183-224.

LACHIVER, Marcel (1982), *Vin, vigne et vignerons en région parisienne du XVIIe au XIXe siècles*, Pontoise, Société historique et archéologique de Pontoise, du Val-d'Oise et du Vexin, 957 p.

LACHIVER, Marcel (1988), *Vins, vignes et vignerons. Histoire du vignoble français*, Paris, Fayard, 714 p.

LAURENT, Robert (1958), *Les Vignerons de la Côte d'Or au XIXe siècle*, Paris, Belles Lettres, 572 et 284 p.

LE ROY LADURIE, Emmanuel (1966), *Les Paysans du Languedoc*, Paris, SEVPEN, 1966, 2 vol., 1035 p.

LE ROY LADURIE, Emmanuel (1977), 'Les Masses profondes : la paysannerie', in Fernand BRAUDEL, Ernest LABROUSSE (eds), *Histoire économique et sociale de la France*, vol. I, Paris, Presses Universitaires de France, 8-1029 p.

LORENZ, Joseph R. (1866), *Die Bodencultur-Verhältnisse des österreichischen Kaiserstaates*, Wien, Seidl, XI-368 p.

LOUBERE, Leo A. (1978), *The Red and the White. A History of Wine in France and Italy in the 19th Century*, Albany, State University of New York Press, XX-401 p.

MENDELS, Franklin (1980), 'Seasons and Regions in Agriculture and Industry during the Process of Industrialization', in Sidney POLLARD (ed.), *Region und Industrialisierung*, Göttingen, Vandenhoeck und Ruprecht, p. 177-195.

MUSSET, Benoît (2008), *Vignobles de Champagne et vins mousseux. Histoire d'un mariage de raison (1650-1830)*, Paris, Fayard, 789 p.

NETTING, Robert McC. (1993), *Smallholders, Householders. Farm Families and the Ecology of Intensive, Sustainable Agriculture*, Stanford, Ca., Stanford University Press, XXI-389 p.

PARENTI, Guiseppe (1939), *Prime ricerche sulla rivoluzione dei prezzi in Firenze*, Firenze, Carlo cya, Publicazioni della R. Università degli studi di Firenze, 124 p.

PIJASSOU, René (1980), *Un Grand vignoble de qualité : le Médoc*, Paris, Tallandier, 2 vol., XXIV-1473 p.

PINTO, Giuliano (1982), *La Toscana nel tardo medio evo. Ambiente, economia rurale, società*, Firenze, G.C. Sansoni, VI-502 p.

PRIBRAM, Alfred Francis (1938), *Materialien zur Geschichte der Preise und Löhne in Österreich*, Wien, C. Ueberreuters Verlag, XXXVIII-879 p.

RAPETTI, Vittorio (1984), *Uomini, collina e vigneto in Piemonte, da metà ottocento agli anni trenta*, Alessandria, Ed. dell'Orso, 342 p.

SCHULZ, Knut (1985), *Handwerksgesellen und Lohnarbeiter*, Sigmaringen, Thorbecke, 477 p.

SERENI, Emilio (1981), 'Per la storia delle piú antiche tecniche e della nomenclatura della vite e del vino in Italia', in Emilio SERENI (ed.), *Terra nuova e buoi rossi, e altri saggi per una storia dell'agricoltura europea*, Torino, Einaudi, p. 101-214.

SERENI, Emilio (1997), *History of the Italian Agricultural Landscape*, Princeton, Princeton University Press, XLIV, 389 p.

SIMONIN, Jean-Pascal (1980), 'Une Interprétation de l'effet Giffen', *Revue Économique*, 31, p. 347-362.

SIMONIN, Jean-Pascal (1992), 'La Crise d'Ancien Régime : un essai de justification théorique, *Histoire & Mesure* 7/3-4, p. 231-247.

VAN DER WEE, Hermann (1963), *The Growth of the Antwerp Market and the European Economy*, The Hague, Nijhoff, 3 vol. LIX-562 p., XI-435 p., 168 p.

VILAR, Pierre (1953), 'Géographie et histoire statistique. Histoire sociale et techniques de production. Quelques points d'histoire de la viticulture méditerranéenne', in *Éventail de l'histoire vivant. Hommage à Lucien Febvre*, Paris, A. Colin, vol. I, p. 121-135.

VILAR, Pierre (1962), *La Catalogne dans l'Espagne moderne*, Paris, SEVPEN, 3 vol. 717, 586, 570-[116] p.

VOLK, Otto (1993) 'Weinbau und Weinabsatz im späten Mittelalter. Forschungsstand und Prespektiven', in Alois GERLICH (ed.), *Weinbau, Weinhandel und Weinkultur*, Stuttgart, F. Steiner, p. 49-164.

WEIGL, Andreas (2003), 'Frühneuzeitliches Bevölkerungswachstum', in Peter CSENDES & Ferdinand OPLL (eds), *Wien. Geschichte einer Stadt*, Wien, Böhlau, vol. II, p. 109-131.

ZUMKELLER, Dominique (1992), *Le Paysan et la terre. Agriculture et structure agraire à Genève au XVIIIe siècle*, Genève, Éd. Passé présent, 375 p.

14. The rapid expansion of vine-growing in the Epernay vineyards, 1650-1830: the necessary creation of a vine-growing proletariat

Benoît Musset

I. Introduction

Historians have always stressed the early specialisation in vine-growing within the traditional rural economy (Dion, 1959)[1]. Roger Dion's work (Dion, 1952: 205-226) made clear the link between wine-consuming urban centres and such specialisation, but since then the implications of this connection have not been fully explored. Indeed, historians have almost entirely been occupied by the technical aspects of early quality vine-growing rather than with its specifically economic and social aspects; the only notable exception to this general preoccupation with the history of the major estates and the birth of quality wines is Marcel Lachiver's monograph on vine-growing in the Paris region and his general work on the history of French wines, where he clearly sets out how specialisation in vine-growing came about (Lachiver, 1982 and 1988).

In the 1650s the process of specialisation was already well advanced in the Marne Valley around Epernay. This zone lay at the heart of the vineyard called the *vignoble de Champagne*, a term which at that time encompassed the Montagne de Reims and the Marne Valley, but excluded all other vineyards in the province of Champagne (Musset, 2006). In 1697, in his *Mémoires pour l'instruction du duc de Bourgogne,* Larcher, the intendant of Champagne, wrote that 'very little grain is gathered in this taxation district (élection), not enough to feed its inhabitants, who have to get it from the surrounding region. Its principal treasure is rather its wines, which are exceedingly good everywhere, especially those produced in Hautvillers and the valley of Pierry, Cumières, Ay and Mareuil' (Brancourt, 1983: 236).

---

[1] Apart from market gardening, other specialisations, in cattle breeding or fruit farming, were generally taken up at a later date.

Agricultural specialisation and rural patterns of development, ed. by Annie Antoine, Turnhout, 2016 (*Rural History in Europe*, 12), p. 273-287.

BREPOLS PUBLISHERS DOI 10.1484/M.RURHE-EB.5.112272

This particular production area has been labelled *vignoble de rivière*, *vignoble* d'Epernay or even *vignoble* d'Ay (Musset, 2006: 55-159) ever since the seventeenth century. Indeed, around the town of Epernay, which had barely more than 5,000 inhabitants, there were nearly a dozen villages given over almost exclusively to vine-growing, in an area stretching about fifteen kilometres from east to west on both flanks of the valley: Mareuil-sur-Ay, Ay, Dizy, Champillon, Hautvillers, Cumières and Damery on the right bank; Chouilly, Pierry, Epernay, Mardeuil on the left. Beyond them, on either side of the valley, woodland, fields and meadows made up a landscape of mixed farming where vineyards were less in evidence. There were two other areas of specialized vine-growing in the vignoble de Champagne: the *Montagne de Reims* around Verzy and Sillery and the Avize vineyard to the south of Epernay. This study leaves aside these two zones despite their many resemblances to the Marne valley vineyards, as the documentation for them is rather spotty.

Figure 14.1. Map of Épernay vineyards

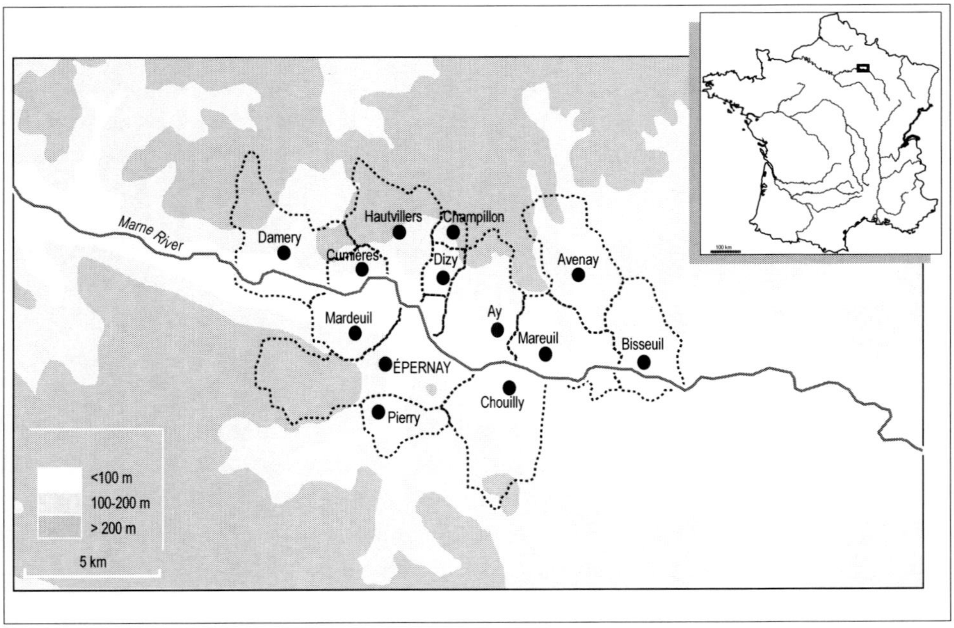

Source. Map by the author.

This paper aims to discover how specialisation in vine-growing generated a particular type of economy and society. This specialisation was encouraged by a commercially-oriented economy and a demand for a broad spectrum of wine quality, ranging all the way from the modest wine intended to quench the thirst of town and country folk to the tête de cuvée and bottled wines destined for aristocratic tables, resulting in a clearly defined hierarchy of production units. On the one hand were absentee owners, with a small number of properties measuring more than 3 hectares, a hefty size for that period and region; on the other were the majority of wine-producing holdings, individually tiny but more numerous and together occupying more space. And yet, the outcome of this contrast was not class struggle but complementarity, for specialisation depended on the complementary nature of these two operating categories: one could not work properly without the other. However, as we shall see, this complementarity changed as the proportion of high-quality production grew, to the point where the relation between the two systems of vine-growing was reversed. First the differing degrees of specialisation in the villages of the Epernay vineyard will be examined. Then the land held by the two types of proprietors – absentee (forain) or bourgeois on the one hand and small vine-growers on the other – will be looked at in more detail, as well the forces for change in both type of holding. A final section will try to explain how the increased power of the big producers and wine-merchants in the course of the eighteenth century gradually made small vine-growers more dependent, transforming them from wine producers into vineyard workers.

II. Highly specialized vine-growing

The degree of specialisation can be measured first of all by examining the surface planted in vines and how it changed and how it was related to other types of farming. It must be admitted that we do not know much about the total area cultivated before the second half of the eighteenth century. Never the less, a few scattered facts give at least some idea of the general trend. During the sixteenth century, the period when the wines produced in Ay started to gain a reputation for themselves with their new wines (vins clairets), a very high level had already been attained. A document from a lawsuit in 1599 claimed Epernay and Mardeuil no longer contained more than 900 arpents (an arpent was a little less half a hectare), and only 600 of them in good condition, compared to 2,000 arpents before the Wars of Religion. Two other documents from lawsuits mentioned areas of, first, 2,000 arpents in 1645 and then 3,000 arpents in 1684[2]. On a smaller scale, the Abbey of Saint-Martin in Epernay having cultivated 20 arpents of vines in 1561, worked no more than 5 in 1599, slowly increasing the number to 9 arpents in 1650 and finally regaining the lost ground in 1715 with 20 arpents (Musset, 2006: 309-310). Hence, we should not assume that this specialisation progressed in a straight line. It underwent significant variations, a sign that it was in competition with other types of farming. There was a good deal of arable land to the east of Epernay, in a zone of transition to the chalk-lands of

[2] This last figure seems excessive as it had not been not been attained by the end of the eighteenth century even though there had been a widespread increase of surface under cultivation (Musset, 2006: 309-310).

Champagne, and also beyond Mareuil-sur-Ay. Then too, the gentle slope at the foot of both sides favoured grain growing, and the marshland at the bottom of the valley provided precious pasture. At the top of the slopes, the large area covered by forests made possible a good deal of complementary activity. In Epernay itself, there were also artisan occupations such as earthenware pottery, tanning and transport trades. First of all came shipping along the Marne by the bargemen of Epernay, Mareuil, Ay and Cumières, who moved not only wines, but also various products from Rheims and upstream: iron from the Haute-Marne and grain from Vitry-le-François and Châlons[3]. Then there was overland transport, with the Paris-Metz road that gave birth to large numbers of taverns and inns serving an extremely active haulage trade, in which the country people around Épernay were deeply involved[4]. Despite the shortage of grain often mentioned at the time, this rather open regional economy was adaptable. It was not predestined to specialize in vine-growing, and it could face up to economic setbacks by turning to other activities[5].

And yet, the fact remains that the eighteenth century was marked by a deep and permanent tendency to specialize in vine-growing. The vineyards of Cumières, almost entirely on the right bank of the Marne, covered 140 *arpents* (59 hectares) in 1700, and 324 (137 hectares) in 1773, according to a reliable land-use survey of the whole region carried out by the intendant's office[6]. In Champillon, a few kilometres further on, a survey (*terrier*) dating from around 1600 showed a vineyard stretching over 93 *arpents* (40 hectares); by 1773, it covered 165 *arpents* (71 hectares)[7]. The celebrated vineyard at nearby Hautvillers is said to have grown from 100 *arpents* (42 hectares) in 1636 to 600 (253 hectares) in 1773[8]. On the other side of the Marne, the vineyard of Pierry grew from 221 *arpents* (96 hectares) in 1738 to 330 (143 hectares) in 1773[9]. Except for the years between 1690 and 1710, which were plagued by recurrent bad weather and witnessed a considerable setback in vineyard area, the general trend is absolutely clear: the vineyards increased in size in the course of the eighteenth century and probably even during the latter half of the seventeenth century[10]. Moreover this happened even though the planting of new vines was officially prohibited – the King's

---

3 Archives Départementales de la Marne, Châlons, C 1743.

4 Archives Nationales, G 2 26, mémoires des directeurs des aides, années 1770.

5 There is no question that it would highly interesting to examine the fluctuations of the local economy in the sixteenth and seventeenth century from various angles. Notarial records are quite well-preserved, as well as the invaluable Treasury registers of Châlons, containing a wide range of information on the economy from 1580 to 1630. The town records of Epernay also provide much useful material.

6 Archives Départementales de la Marne, Châlons, C 430.

7 Archives Départementales de la Marne, Châlons, H 1312; C 430.

8 Bibliothèque Municipale d'Epernay, fonds Chandon de Briailles, liasse 261, piece n°16, letter of 16 October 1821.

9 Archives Départementales de la Marne, Châlons, H 1178; C 430.

10 These conclusions are based on data concerning the Reims vineyard alone (Musset, 2006: 217-222).

Council passed a decree to this effect in 1729, reinforced by a second one in 1731 applying to the whole kingdom (Musset, 2006: 217-222). In 1732, the intendant Le Pelletier de Beaupré took a dim view of these changes:

> First of all, this abundance. Secondly, it accustoms the peasants to debauchery and drunkenness. Finally, vine growing keeps them from cultivating grain, which would be far more advantageous than this useless abundance of wine.[11]

Table 14.1. Percentage of area occupied by vines in the villages of the Epernay region, 1773-1837

	1773		*1837*	
Right bank	*Hectares*	*%*	*Hectares*	*%*
Avenay	198		-	-
Mareuil	68	6	-	-
Ay	305	29	-	-
Dizy	117	36	-	-
Champillon	71	49	-	-
Hautvillers	253	21	-	-
Cumières	137	46	-	-
Damery	447	29	392	25
Left bank				
Bisseuil	71	7	-	-
Chouilly	81	5	246	15
Épernay	309	14	328	14
Pierry	143	28	111	22
Mardeuil	167	18	203	22

Source. MUSSET (2008).

The 1773 survey makes it possible to compare the proportion of the soil occupied by vines in the villages examined here. The proportions might not look very impressive at first glance, as since they never exceed 50 per cent of village land. However, if we realize that about one-third to one-half of this land was covered by woods and forests (38 per cent at Epernay, 40 per cent in Dammery) and that this proportion seems to have changed very little over the years, we can see that vineyards usually took up between half and two thirds of cultivated land. Thus in a woodless village like Cumières, vineyards account for 46 per cent of land, the other half being made up almost entirely of marshy meadows on the edge of the Marne. By contrast, the villages of Bisseuil and Chouilly mostly opened onto the chalk country of Champagne and

[11] Archives Départementales de la Marne, Châlons, F 148, f° 16.

had few sloping valley sides. This explains the sparseness of vines in these regions despite the reasonably good reputation of their product. Not surprisingly, increasing the area planted in vines pushed up the relative share of land occupied by vines. In the three villages where we have data that allows us to go back into the seventeenth century, we know that the area covered by vines increased from 27 to 49 per cent in Champillon (1600-1773), from 4 to 21 per cent in Hautvillers (1636-1773) and from 2 to 5 per cent in Chouilly (1614-1773); the sharp rise in the share of total surface under vines in Cumières between 1700 and 1773, from 20 to 46 per cent of total acreage, is also worthy of note.

As often happens, it is harder to find sources for the first third of the nineteenth century. A fragmentary survey of 1837 covering only a few villages, reveals that the increase in vine-growing stabilized, except in Chouilly, where the vineyards moved slightly down the slope of the outlying hill of Saran towards the lower zones[12].

Nevertheless, the most significant impact of wine-growing was on the local social structure. The study of this aspect is hampered by fragmentary records. Direct tax records for the *taille tarifée* in the Epernay district, were not kept very well, unlike those at Reims. Those from the 1780s do not always mention professions. Many were used in several successive years, resulting in numerous deletions; names and figures were crossed out to the extent that the documents became illegible. This study had to be limited to a small number of lists. In Ay, where 29 per cent of the land area was used for vine-growing, 55 per cent of those paying the taille were designated as vine-growers (vignerons) and 3 per cent as cottagers (manouvriers). And yet, we are talking about a small town with flourishing rural industries. In Mareuil-sur-Ay, only 6 per cent of the area was occupied by vineyards, but 31 per cent of family heads were vignerons, 31 per cent cottagers and only 6 per cent landowning farmers (laboureurs). In these two places winegrowing generated respectively 12 and 6 jobs for master coopers (2 and 4 per cent of the family heads)[13]. The village of Fleury-la-Rivière, next to Hautvillers, but attached to the Reims tax district (élection) and therefore with better-kept rolls, occupied 167 hectares in 1773; no less than 71 per cent of its family heads were vignerons in 1742 and 78 per cent in 1773[14]. This social structure

12 Archives Départementales de la Marne, Châlons, 146 M 14. The areas are almost identical today. The vineyards of Epernay shrank slightly because of the urbanisation of the eastward-facing slope that dominates the city; only 10 per cent of the land was under vines by 2000. By contrast, vineyards continued to spread out over the arable in Chouilly (31 per cent vines), Bisseuil (12 per cent) and Mareuil (24 per cent) under the pressure of a property tax, which is about to reach unheard-of levels. The Champagne region holds the world record at the moment for the highest price for vines per hectare.

13 In the *Montagne de Reims* (Rheims district), where sources are explicit, 70 per cent of the working population were vine-growers in Verzy and Verzenay in the 1740s and again in the 1770s.

14 These proportions, ranging from one half to three quarters can be found in the highly specialized villages of the Loire Valley, for instance; in the 16 rural parishes of the Amboise area, the proportion of vine-

dominated by vine-growers not surprisingly carried on through the first third of the nineteenth century. However, the term vigneron gradually disappeared from the tax rolls to be replaced by the more general expression cultivateur, disappointingly less specific for the historian. For the same reason, property tax registers are short on useful information. As for the cadasters, which were drawn up for the most part in the early 1830s, the amount of work needed to extract useful information from these documents has made it impossible to use them for the time being. But, if we start from the figures at the end of the Old Regime, it does not seem outrageous to estimate that vinegrowers made up from 50 to 80 per cent of the total working population in the wine-producing villages.

III. Exclusive vine-growing: a specialisation by default

The homogeneity of the population around Epernay involved in vine-growing was at any rate perfectly evident to Jean-Baptiste Mennesson, one of the most knowledgeable observers of the vineyards, who in 1806 ventured the expression 'a people of the vineyards':

> The people in this region are comfortable, but none are excessively wealthy; you don't find distressing contrasts between extreme opulence and extreme poverty; there is practically no begging. Anyone who puts his arms and heart into his work can be sure of earning his daily bread; he can bring up his family by the fruits of his labour and put money aside for his old age.... In a wine-producing region good health and two arms is all you need to be sure of employment and your daily bread. The profession of winegrower does not require laying out a lot of capital, and a thrifty worker, who saves some of what he earns by selling his labour to other farmers, can thereby indulge in the hope of becoming himself the owner of a few vines (Mennesson, 1806: 86-87, 92).

Nevertheless, the same author makes a point of distinguishing between different categories of winegrowers:

> The class of the farmer-vinegrowers (cultivateurs-vignerons) can be subdivided into two others, the landowners and the cottagers (manoeuvriers): the situation of the former group is more comfortable in many respects, it is the natural result of their wealth... Cottager-vinegrowers enjoy the advantages of the former group only to the extent that their abilities or personal means allow. Their situation always remains a bit precarious, depending on good or bad years (Mennesson, 1806: 88-89).

growers came to 66 per cent of all households, with figures running between 51 and 72 per cent (Jollet, 1997: 69-88). In Coulanges-la-Vineuse, an important market town of the Auxerre vineyard and very much like Ay, vine-growers account for 48 per cent of the population subject to the *taille* (Desaive, 1996: 83-94).

Finally, Mennesson mentions a third category, the 'rich landowners' of Ay, the 'well-to-do owners' of Hautvillers and the big landowners of Cumières who 'only maintain a residence there' (Mennesson, 1806: 110). He might have added the broker/wine merchant/retailer as a fourth type of active participant in this system of specialized production. Their role will be presently described in detail.

Table 14.2. Families, property and wine-growing in the *taille* roll at Ay, 1775

	Heads of family	*% of heads < of family*	*Total surface area planted with vines*	*% of total*	*Average acreage of estates (all possessions combined)*	*% of vines on the estates*
> 3 ha	22	3	94.8	19	9.79	44.00
2 to 3 ha	33	5	75.8	15	5.54	41.44
1,5 to 2 ha	41	6	71.8	15	4.02	43.61
1 to 1,5 ha	93	14	110.9	22	1.98	60.26
0,5 to 1 ha	131	20	92.9	19	1.03	69.02
0 to 0,5 ha	191	30	48.6	10	0.38	67.59
0	132	21	0.0	0	0.11	0.00
Total	643	100	495,1	100	1.51	51.11
Average			0,96			

Source. Musset (2008).

This description of the situation in the early nineteenth century was valid for at least the previous century. Three levels of holding were beginning to emerge in the course of the eighteenth century, which was the time when specialisation in wine-growing became more and more common. The taille roll for Ay in 1775 gives a good general view of the hierarchy (see Table 14.2).

This document takes us into the very heart of specialized vine-growing, even though the overall picture is to some extent distorted by the over-representation of artisan family heads and the absence of the noble landowners, who were exempt from the taille, and therefore not recorded here. At the top were the owners of the big holdings. To judge from the documents assembled in my thesis research, it was in the course of the seventeenth century that large commercially-oriented holdings emerged. Thus, the estates of several religious establishments expanded considerably between the end of the sixteenth and the first third of the eighteenth century, more by purchase than planting, unlike what happened in the Medoc at the same period

(Pijassou, 1980). There had been some biggish religious, absentee and bourgeois properties ever since the early Middle Ages, but these tended to aim for the domestic market (Devroey, 1989: 154). It is true that a great deal of wine had been sold outside the region since the thirteenth century (Devroey, 1989: 143-144). However, as we have shown, there was hardly any difference between the production methods on different types of holdings until the latter half of the seventeenth century. Minimal labour was expended on the vines; such differences as emerged were owing to the intrinsic quality of the grapes and the vineyard (Musset, 2006: 506-509).

The threshold of size that set apart the large holdings from ordinary bourgeois holdings producing for the domestic market, lay at around three hectares. With an average long-term output of 20 hectolitres per hectare, these big estates could produce a yearly average of 30 casks, each containing approximately 200 litres of wine. Alternately they might make about ten casks of top quality wine (*tête de cuvée*), produced with care and according to the new wine-making methods, usually fetching 100 to 250 livres for the best casks. Below the 3-hectare threshold, the unreliability of harvests and the heavy investment needed to make white or gris wine by the new methods, whether in casks or bottles, made it almost impossible to go into the trade in 'River wines' (*vins de Rivière*). In the neighbouring villages of Epernay no more than a hundred owners belonged to this group of big bourgeois vine-growers. A closer look at the situation in Ay shows Paris burgesses (*bourgeois de Paris*), officials from Châlons and Epernay, wine merchants (several of them listed among the producers of bottled wines) and also a few big vine-growers like Louis François dit Chauffour whose wealth and success in business on a large scale has been the subject of earlier study (Musset, 2006: 619-623).

There is a very clear link between the size of a property and strong diversification. In Ay itself, vines accounted only for about 40 per cent of the area held by leading wine-producers. Nor was vine-growing their sole activity; in addition to 3 to 5 hectares of vines or more, we generally find woods, cultivated land and, especially, valuable meadows. Whenever post-mortem inventories allow more detailed examination, it appears as a rule that these people had big farms in the chalk part of Champagne, substantial wooded areas, and vines in other villages. For the big landowners vine-growing was part of a complex framework of highly diversified agricultural and financial activity. Take for example the owner and wine merchant Lasnier, whose family was long settled in Ay. Upon his death in 1796 he left 12.4 hectares of vines, including 5.3 in Ay, but his vines accounted for only 12 per cent of his landed capital, whereas 73 per cent was made up of various farms in the chalk part of Champagne and in the Ardennes[15].

---

[15] Archives Départementales de la Marne, Châlons, 4 E 4053.

By contrast, the farther down the hierarchy, the more vine-growing tended to be the sole activity. The vineyard was basically the only land that small vine-growers worked, accounting for more than two thirds of the inheritances of people cultivating less than one hectare (Musset, 2006: 549-557). With an average yield of 20 to 25 hectolitres per hectare, it seems clear that most vine-growers did not usually produce more than ten casks of wine a year; only half of that could be sold at a decent price, between 50 and 75 livres for red or white wine. And just consider the situation of the holdings of less than half a hectare that made up 10 per cent of all the vineyards in Ay, and whose total harvest came to only four or five casks a year. Taking into account the relative absence of specialized equipment (a subject we do not have space to examine here), and after their own, inevitable, home consumption, they could have put only two or three casks of wine on sale, for a price that ran from 30 to 35 livres a cask in a normal year. Even though household heads described as vine-growers operated 61 per cent of the Ay vineyards, this group paid out 3,470 livres a year in net loan interest owed on a principal amounting to nearly 70,000 livres. They owned practically no woodlands or meadows.

Make no mistake: the majority of vine-growers did not choose specialisation; they were forced into it. Except for a very small group of a few vine-growers in each village – usually four or five in the *Montagne de Reims* and about twenty in the wine-producing village of Ay – winegrowers' land-holdings specialized in a rough and fragile form of vine-growing, capable of producing 35 to 40 hectolitres a hectare one year and 10 to 12 in the next. By contrast, the big owners diversified their activity, more easily avoiding risks by juggling the many resources offered in both the rural and urban worlds. Beyond the group of small and vulnerable vinegrowers tempted by the high yields of a less meticulous viticulture, specialisation in vine-growing was not on offer as a positive strategy at the end of the *Ancien Régime* any more that it was available in the 1820s, where the situation had hardly changed in this regard (Musset, 2006: 980-989). Even though the wine market, with its urban and rural consumers, gave the vineyards a great impetus in the course of the eighteenth century, it generated forced specialisation at the bottom rather than at the top of the production hierarchy. Seen from this angle, the call of the market appears to have generated an unexpected drive to consolidate a social and economic system based on a multitude of small growers to whom, despite their feeble means of production, it furnished market opportunities.

IV. Between complementarity and dependency. Specialisation as a system

This fragility of the vine-growers' world and their remarkable hold over a substantial share of the land was no recent discovery, but I was able to estimate that it extended to the equivalent of two-thirds to three quarters of the land surface depending on the nature of the vineyard (Musset, 2006: 469-473). This hold on the land has been shown to exist in many regions, to such an extent that this world of vine-growers has been deemed an embryo of an egalitarian rural society resistant, for good or ill depending on the observer's ideological standpoint, to takeover by the great landlords. And yet, if we look at the matter from a different perspective, we might reconsider, and see in this intertwining of cautiously diversified big holdings and dangerously specialized small ones the essence of an overall system that, to a certain extent, favoured both parties.

Vine-growing was a risky business requiring regular persistence in the numerous operations needed to maintain the vines, so bourgeois owners needed a large workforce. By contrast small vine-growers could not depend for a living on their own micro-plantations alone. Many of them had to look for complementary financial resources: keeping a cow, hiring out a donkey for various cartage jobs, collecting vine-shoots after pruning, exercising their rights to use the woods and meadows, or fraudulently selling wine by the jar. However, it was above all by paid work on the big estates that the vine-growers secured some indispensable regularity in their income.[16] In the 1740s, the wine operation at Pierry, owned by the monks of Saint-Pierre-aux-Monts of Châlons, hired labourers directly for an average of 196 days per arpent of vines, paying between 6 and 20 sols a day. Half of their property was worked this way. Vines further away and more difficult to watch over were cultivated by piece-workers paid by the task. The cost of all vineyard work, except for grape picking and the tough work of pruning the vines, came to around 40 *livres* a year for each arpent. Absentee owners had to spend at least 60 to 70 *livres* a year on each cultivated arpent of vines, taking into account the cost of the grape harvest, the grape pressing, the cartage and the wine-making process itself; the monks of Pierry paid as much as 94 *livres* an arpent during the period between 1737 and 1749.[17]

---

[16] On this subject I do not fully agree with Jean-Paul Desaive who, writing of the Auxerre vineyard in the eighteenth century, interprets the increasing number of salaried vine-growers as above all a drop in status and a sign of weakness. This is undeniable, but it is also an element of consolidation, which seems to be part and parcel of the vine-growing activity in an ecological *Ancien Régime* heavily dependent on climatic conditions. After all, the big owners themselves were not specialized and their diversified income ensured that they would have something on which to fall back (Desaive, 1996: 94).

[17] Archives Départementales de la Marne, Châlons, H 925.

In the light of these figures, we can understand what Mennesson meant when he said that 'good health and arms' were the best insurance against poverty in the vineyard. The collectors of direct taxes were probably not wrong in calculating that each vine-grower gained 40 *livres* from his 'industry', that is, his earned wages, using a downward-sloping scale of charges according to age. The most active families, by dint of hard labour, must have managed to get a salaried income of 100 or 120 *livres* per annum, at least as much as what they could hope to get by the sale of their wine. Yet this did not prevent the vine-growers from pouncing on the smallest plots of vines on sale in order to enlarge their holdings. Towards the end of the *Ancien Régime* and even into the 1820s, they were actually the leading figures in the wine property market (Musset, 2006: 980-984).

Reinforced specialisation in eighteenth-century vine-growing led to a fairly robust system: the vine-growers remained in control of most of the vineyards producing ordinary wines for the local market or even – in the case of their best vintages – for Paris, Picardy or Flanders. Their tiny holdings gave them a not insignificant, though variable, income. As for the big owners, they built up relatively large estates, but beneath a threshold of about 10 hectares, except in a few cases (Musset, 2006: 491-506).

Their wines reached record sale prices, but production expenses also shot up. The cost of maintaining vines had increased tremendously as did the outlay on maintenance once they made their precocious entry into the era of wine-bottling. For the big proprietors, a wide range of agricultural activity and a great deal of investment in *rentes* made up for the variable harvests. A wholesale increase in the area under vines was impossible to any significant extent because it was so expensive to put each additional arpent of vines into cultivation. In other words, growing specialisation was based on a sort of equilibrium between the big diversified owners on the one hand and a multitude of small vine-growers on the other. The latter group were extremely specialized and above all highly proletarianized, insofar as their earned income was essential to balance their budget.

According to my calculations for the end of the *Ancien Régime*, based on the estimations produced for the lists of grievances (cahiers de doléances) drawn up in Ay, for the Estates General of 1789, an arpent of vines, which was the extent of the typical holding of a small specialized vine-grower, with an exceptional vintage (8 casks per arpent priced at 54 *livres* a cask) could bring in an income of 297 *livres* after expenses. In a more ordinary year with a more typical yield of 5 casks per arpent, selling at the standard price (48 *livres* a cask) observed at this level of yield, net income dropped to 32 *livres*. In a disastrous year as occurred between 1767 and 1771, when yields fell to 2 casks an arpent, the vinegrower ran a loss of 39 *livres*, even without having to pay their own labour costs (Musset, 2006: 602). As families spent around 100 to 150 *livres* a year on food grains at the same period, we can better comprehend the extent to which salaried work really formed the great pillar on which this specialized wine economy was built. Without high quality viticulture, high added value and strong demand for manual labour, specialisation in vine-growing, such as existed in the eighteenth century around Epernay, would have been impossible. Most of the small holdings focused on vine-growing were not viable. As a strictly wine-making operation, the Epernay area could never have produced 60,000 bottles of rare and expensive sparkling wines a year in the 1720s, 300,000 in the 1780s and 3 million in the 1800s without the abundant harvests of red or white wines – about 100,000 casks of wine in an ordinary year for the Epernay taxation district – produced by rudimentary vineyards and aimed at a popular market.

Within the chronological limits of my thesis research, the irruption of the wine merchants into this production system from the 1790s and 1800s onwards hardly changed the situation before 1830 at the earliest. It took place during a time when the market for bottled wines was gaining volume, making supply management more and more fragile, especially with a clientele demanding very specific types of wines whose production required ever more complex techniques. The wine merchants did not invest in the property market, preferring to purchase wines which had already been bottled by the big landowners or wines in casks that they themselves blended, or even baskets of grapes at harvest time. Jean-Remy Moët left 23 hectares of wine-growing property when he retired from business in 1832, about the same size as the holdings of the Abbey of Hautvillers before the Revolution, but the other wine merchants contented themselves with holdings of a size comparable to those of the bourgeois, fitting themselves into a stable system of divided roles and risks, which had proved effective for at least one and a half centuries (Musset, 2006: 514-551). The wine trade played the leading role in the whole wine economy of Epernay and its surrounding small towns, offering new opportunities for salaried labour, in the production and handling of wines for example, but it did not transform the production system of specialized vine-growing described here.

Already well advanced in the sixteenth century, if not earlier, specialisation in vine-growing around Epernay went through a period of definite expansion in the seventeenth and eighteenth centuries with an increase in both the area planted with vines and in the number of vine-growers. In July 1789, Arthur Young described a region where 'two-thirds of all the country around, about Ay, Cumiere [Cumières], Piery [Pierry], Disy [Dizy], Hautvilliers [Hautvilliers], &c &c. under vines; and here all the famous Champagne wines are made' (Young, 1794: II, 9). With his obsessive search for enlightened owners of large estates and often appalled by the archaic ways of French peasants, Young naturally emphasized that 'small growers around here are generally very poor and many of them are ruined, because they cannot wait for a favourable price' (Young, 1931: 701). He was not entirely wrong, but he had failed to notice what Mennesson, a man of the vineyards, pointed out about twenty years later, namely that the ordinary vine-grower from Epernay or the surrounding villages was above all a vineyard labourer, a hired hand, relying on wages earned on absentees' holdings in order to keep his specialized micro-holding in operation. A kind of equilibrium had been attained, where two forms of wine-growing found themselves side by side without competing with one another – at least not yet. On the one hand the producers of quality wines, whether in bottles or in casks, aimed at the export market, controlled a quarter to a third of the vineyards; on the other hand growers and small landowners provided very ordinary wines directed to the local market or nearby regions. People in the first group needed the labour force and know-how of the second group; the latter in return required the money of the former in order to balance their household budgets. This interdependence, that brought about a certain balance between the two winegrowing systems, was broken when the merchants came to need more and more grapes to satisfy the strong demand for sparkling wines, of which there were 10 million bottles produced in the 1840s and 30 million in the 1860s. The concomitant collapse of the red wine market which reduced the output of the vine-growers to the absolute minimum, showed up the situation in its true colours, Many continued to produce red wines for local markets while the most substantial owners managed to catapult themselves into the great adventure of bottled wines. By the end of the nineteenth century, even before the phylloxera crisis, the vine-growers had been only too evidently reduced to simple vineyard workers serving a wine business dominated by bottled wines. In 1911, the vine-growers of Ay unfurled red flags and marched to the tune of the *Internationale* (Bonal, 1984: 161). It was mainly after the 1920s that the cooperative movement gradually made the vine-growers once again wine producers with a central role in a specialized wine region.

Bibliography

BONAL, François (1984), *Le Livre d'or du champagne,* Lausanne, éd. Du Grand Pont.

BRANCOURT, Jean-Pierre (1983), *L'Intendance de Champagne à la fin du XVIIe siècle. Édition critique des Mémoires pour l'instruction du duc de Bourgogne,* (see : 'Mémoire de Larcher'), Paris, éd. du CNRS.

DESAIVE, Jean-Paul (1996), 'Vignerons urbains et vignerons ruraux du comté d'Auxerre sous l'Ancien Régime', *Bulletin du Centre Pierre Léon d'histoire économique et sociale,* 3-4, p. 83-94.

DEVROEY, Jean-Pierre (1989), *L'Éclair d'un bonheur : une histoire de la vigne en Champagne,* Paris, La Manufacture.

DION, Roger (1952), 'Querelle des anciens et des modernes sur les facteurs de la qualité du vin', *Annales de géographie,* LXI, november-december, reproduced in *Le Paysage et la vigne*, Paris, Payot, 1990.

DION, Roger (1959), *Histoire de la vigne et du vin en France, des origines au XIXe siècle*, Paris, Flammarion, 768 p.

ÉTIENNE, Michel (1994), *Veuve Clicquot Ponsardin. Aux origines d'un grand vin de Champagne,* Paris, Economica.

JOLLET, Anne (1997), 'Les Vignerons et le marché immobilier de l'Ancien Régime à l'Empire, dans la région d'Amboise', *Dix-Huitième Siècle,* 29, p. 69-88.

LACHIVER, Marcel (1982), *Vin, vignes et vignerons en région parisienne du XVIIe au XIXe siècles,* Pontoise, Société historique et archéologique de Pontoise, du Val-d'Oise et du Vexin, 957 p.

LACHIVER, Marcel (1988), *Vins, vignes et vignerons. Histoire du vignoble français,* Paris, Fayard, 714 p.

MENNESSON, Jean-Baptiste-Antoine (1806), *L'Observateur rural de la Marne ou Tableau historique et topographique des vignobles de ce département,* Épernay, Warin-Thierry.

MUSSET, Benoît (2006), *Le Vignoble de Champagne, de la naissance des vins mousseux à celle des maisons de champagne (1650-1830). Les transformations d'un univers vinicole, social et commercial*, unpublished doctoral thesis, University of Reims, 1139 p.

MUSSET, Benoît (2008), *Vignobles de Champagne et vins mousseux,* Paris Fayard, 789 p.

PIJASSOU, René (1980), *Un Grand vignoble de qualité : le Médoc,* Paris, Tallandier, 1743 p.

YOUNG, Arthur (1794), *Travels during the Years 1787, 1788, and 1789. Undertaken more Particularly with a View of Ascertaining the Cultivation, Wealth, Resources, and National Prosperity of the Kingdom of France*, Second edition, London, W. Richardson, 1794.

YOUNG, Arthur (1931), *Voyages en France en 1787, 1788 et 1789,* éd. by Henri Sée, 3 vols, Paris, Armand Colin.

15. Hop-farming specialisation in seventeenth-century Swedish rural society

Pia NILSSON

I. Introduction

The main focus of this paper is to discuss hop-farming as one aspect of early specialization in seventeenth century rural Sweden. The regions studied are the provinces of Västergötland in south-western Sweden and Uppland in central Sweden (Figure 15.1). The main sources are geometrical maps, and the paper also includes a discussion of the source value of the maps. The agrarian system in the area under consideration combined grain growing and animal husbandry with other mixed activities. These pursuits differed according to geographical conditions, but included fishing, nut and birch-bark collecting and milling. Production was for subsistence; the ordinary farm comprised two to ten acres and a couple of cows, draught animals (mainly oxen) and some small stock (normally sheep or goats, sometimes pigs), and, according to the size of the holding, access to woodlands, fishing water and grazing rights outside the enclosed area of the hamlet or farm. The arable land was either sown yearly, or divided into a two- or three-field system. In Uppland the two-field system predominated, while the three-field system was common in Västergötland. The farmsteads in the villages were worked on an open field system. The livestock (cattle, sheep and goats) was to a large extent sent to graze in the woodlands outside the hamlet-area for substantial parts of the year. Only the draught animals, milk cows, and young animals were kept in enclosures near the farmstead.

Although subsistence farming was the prevailing system, we know that the early seventeenth century was a period of change at several levels. Among the transformations were the beginnings of individualization and specialization in agrarian society. A possible way to study early specialization is to analyse the prevalence of cash-crops, and I have chosen to study hop farming. Hops were used in the household only to a limited degree, and are useful to analyze as a cash-crop where the main purpose was to sell the produce.

Agricultural specialisation and rural patterns of development, ed. by Annie ANTOINE, Turnhout, 2016 (*Rural History in Europe*, 12), p. 289-303.

BREPOLS PUBLISHERS DOI 10.1484/M.RURHE-EB.5.112273

Hops helped to preserve beer, and beer could at this time be considered one of the most important sources of human energy. Some one-third of the daily energy needs have been estimated to come from the two to three litres of beer that were drunk daily. Brewing was also an important way of saving harvests that otherwise would have been completely destroyed by rain, because even field-germinated barley can be used. In bad times, the bines could be fed to the animals, and the small shoots, similar to asparagus, could be eaten by humans (Karlsson Strese, 2008a: 33-35).

Figure 15.1. Overview of Sweden with the studied provinces Uppland and Västergötland roughly marked

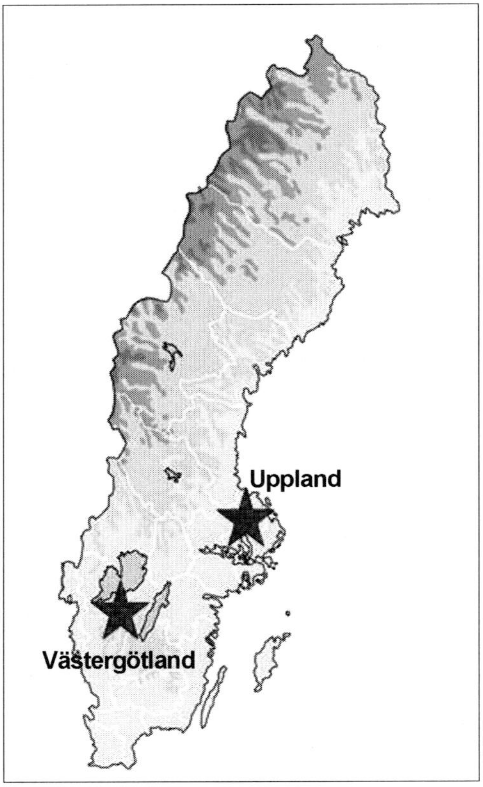

Source. Map by the author.

I consider specialization to be production for the obvious purpose of selling all or most of the produce. I do not mean the saleable or exchangeable surplus that may come from a large grain-harvest due to occasional favourable weather conditions, or an unusual number of surviving lambs or a successful fishing season. The main purpose in these cases is subsistence farming, characterized by a system of varied activities. Subsistence farming includes selling and exchanging products between neighbours and nearby villages, but this multi-functional system must not be confused with specialization.

I have chosen to study hop-farming. The reason for this choice is that the hop was valuable and important, in fact it was the only crop mentioned in Swedish law. During the fifteenth century the law ordered that every farm must grow forty hop-poles. In the seventeenth century the amount had risen to 200 poles per farm (Sveriges Rikes lag 1734, [The Swedish Law] published 1984; Karlsson Strese, 2008: 30). It was obviously considered important that Sweden grew its own hops, and try to limit their importation.

The strong demand for hops made it possible to sell a surplus at a good profit, which made the crop relatively safe as a first step toward specialized farming.

II. Hop-farming according to the geometrical maps and to supplementary sources

II.1. The Swedish geometrical cadastral maps (1630-1655)

The Swedish *geometriska jordeböcker* (geometrical cadastral maps), (1630-1655) give an outstanding view of large parts of rural Sweden. The maps provide the opportunity to view and understand the seventeenth-century rural landscape: the spatial organization of farms and hamlets, farm layout, topographical conditions and, to a certain extent, the vegetation. These seventeenth-century maps are interesting for research because they cover large areas, in a fairly uniform way over a short period of time, twenty-five years.

In the middle of the devastating and costly Thirty Years' War and the emerging agrarian crisis, King Gustav II Adolf established a National Land Survey with a commission to map and describe all Swedish cities, farms and villages including their fields, meadows, forests and other assets of economic importance, together with all mines, harbours, channels and other possible, or future possible, assets. This magnificent ambition was never fulfilled, but about 14,000 maps of single farmsteads and hamlets, with their arable land and hay meadows, were delivered to the Royal Palace between 1633 and 1655 (Baigent, 2005; Tollin, 2005 and Tollin, 2007: 51-57). The scale they used was generally 1:5000. The woodlands and the grazing lands were mentioned, but not usually mapped.

The detailed survey contains a wide range of information on topographical, legal and economic matters. The information was brought together in a text description (*Notarum Explicatio*), annexed to each map. Every single map normally describes one or two farmsteads, or a hamlet. The geometrical cadastral maps were collected in the House of Parliament at the royal palace in Stockholm, and were bound in collections arranged by hamlet, parish and hundred (*härad*).

II.2. The content of the maps

The aim behind the large-scale map project did, of course, affect the choice of information that was considered relevant to the map, and – equally important – what was outside the scope of the task and therefore left out (Dovring, 1953: 27-28). This means that there may have been other resources, but if they were not relevant to the surveyors' task, they were not revealed, or are only indirectly observable. The title "cadastral map" hints at the aim behind the survey. The main objective, although not yet quite understood, was economic rather than fiscal. The crown wanted a better overview of agricultural resources, and it was hoped that this overview could be used to increase production. There were other reasons as well, such as recovering crown land, preventing costly disputes over boundaries, and so forth. (Helmfrid, 1959; Tollin, 1991: 8-16).

The maps show the arable land, the hay-meadows and the farmyard. Ground cover and land use is coherently described, as well as the area and soil quality of the fields. The yield and quality of the hay-meadows is also mentioned. Certain kinds of vegetation and trees were noted. The fences and the farmstead (symbolized by a single house) are displayed. With every map came a short descriptive text, with a standardized content of fiscal statements; surface area, hay yield and a short description of certain other resources. The content may differ slightly according to the region, the year of the survey and the surveyor in charge, but generally the same information is given. Figure 15.2 shows an example of the style and the content of the maps.

Figure 15.2. The style and content of the maps

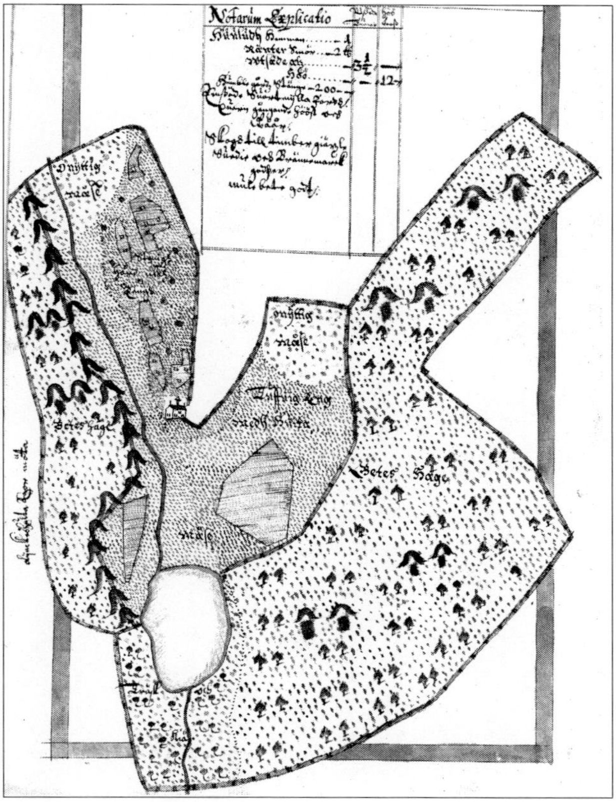

Source. Hulu holding. Öresten: 77. Seglora parish, Marks hundred, Västergötland. Surveyed by Nils Eriksson, 1649-1650.

Note. This map illustrates the appearance and content of the geometrical maps. The arable fields are shown as grey, crosshatched areas and the hay-meadows as green, dotted zones. The grazing areas have basically the same symbol as the meadows, but with more scattered dots. The house symbol represents the farm; the individual buildings are not shown.

II.3. Hop-farming on the maps

Several of the seventeenth century land surveyors describe the extent of the hop-gardens by the number of hop-poles; I therefore assume that the maps will be useful in order to separate the farms that raised hops for subsistence from those that grew them for sale.

The frequency of hop-gardens varied between regions. Certain regions seem to have had none at all, while others had numerous hop-gardens, perhaps four to five per single farmstead. Regardless of the surveyor, the hop-garden was always shown, either by numbers or in the text, as connected to a defined farm, and never as a common resource belonging to the village. The individual surveyors display the hop-gardens in different ways. Some surveyors recorded detailed information about the number of poles; others illustrated the garden with an explicit symbol. (Figure 15.3).

Figure 15.3. A hop farm was private property, never a common resource

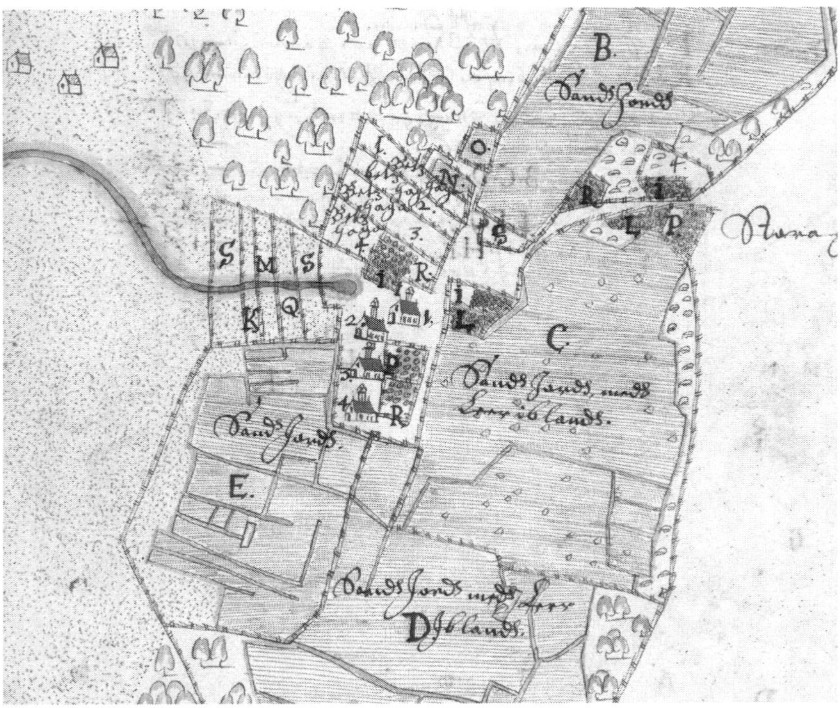

Source. Mehedeby hamlet. A3: 109-110, Tierp parishand hundred, Uppland. 1640-1641.

Note. This map is a good example of land surveyor Sven Månsson's accurate way of showing hop-gardens. The hop-gardens are the green enclosed areas marked by L, R and P. Every hop-garden is connected to a defined farm, and is never a common resource shared by the village (Mehedeby hamlet. A3: 109-110, Tierp parishand hundred, Uppland. 1640-1641).

The instructions that regulated the surveyors' work are kept in the House of Parliament Archive. They mention hops for the first time in 1643: 'even where hop-gardens exist on the farms, as well as their size' (*Äfvenså hvarest humlegårdar finnas till gårdarne, så ock huru stora de äro*) (Ekstrand, 1901: 33). However, the information is present on even earlier maps. There are differences in the frequency with which hop-producing units occur, but they are obviously connected to certain regions and not just to the arbitrary choice of the land surveyor who happened to be in charge in the year of survey.

Figure 15.4a. The hop-farming is concentrated to certain regions

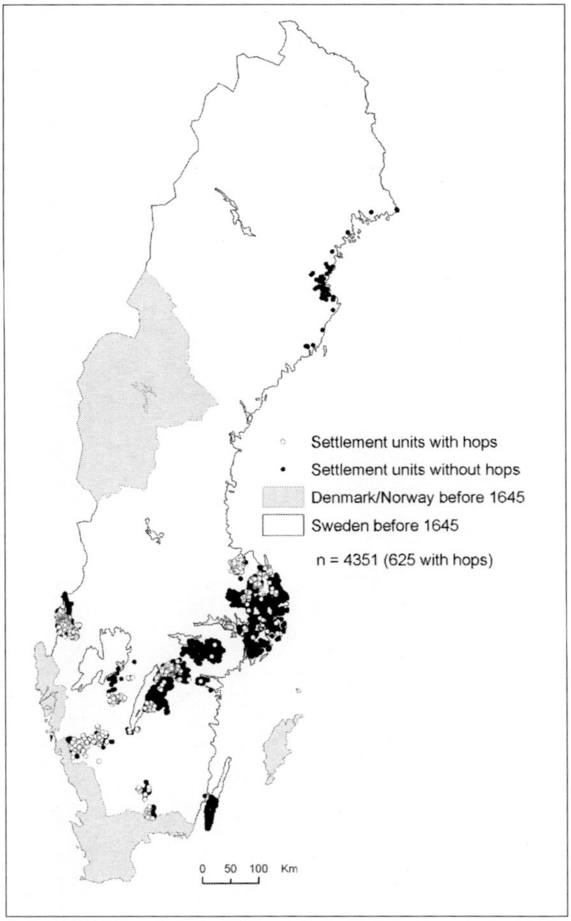

Source. Illustration by Olof Karsvall.

Note. The clusters of settlement units (hamlets or single farmsteads) with hop-gardens. Hops were not grown in all suitable areas, but concentrated in distinguishable regions.

Figure 15.4b. The regions studied provide two different examples of intense hop-growing regions

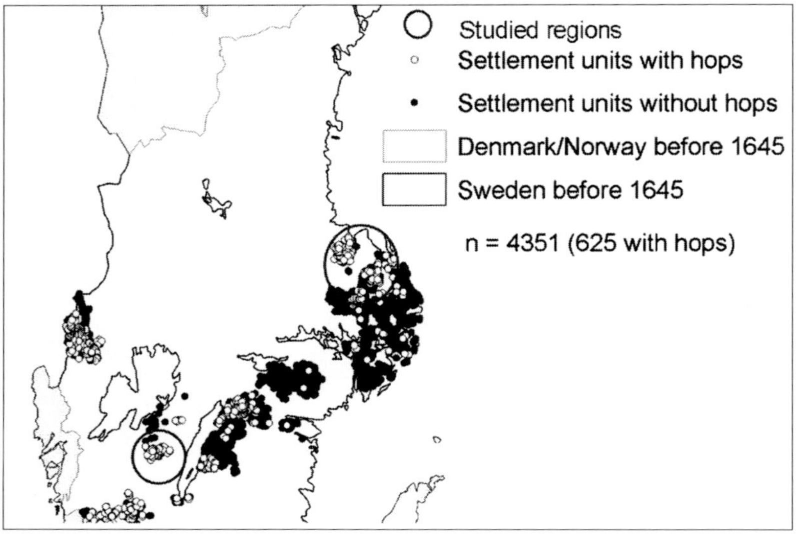

Source. Illustration by Olof Karsvall.

Note. The black circles showing the regions studied within two hop-farming clusters. Illustration by Olof Karsvall.

Assuming the necessary daily consumption of two to three litres of beer per adult is correct, that an ordinary family consisted of two adults and two or three children, and that beer was drunk for most of the year, a family needed some three thousand litres of beer every year. According to genetics and hop-researcher Else-Marie Strese's calculation (Karlsson Strese, 2008b: 47) this means that a family needed about sixty hop-poles, each with two or three plants, for household needs.

The number of hop-producing farms varied significantly between different areas, and distinct concentrations can be estimated (Figure 15.4a). Within the 'clusters' of hop-producing farms (the white circles in Figure 15.4a and Figure 15.4b), the size of the hop-gardens varies considerably, from 30 to 8,000 hop-poles. Before looking more closely into this, we have to decide whether the clusters of hop-producing farms show the true situation; this can be done by comparing the information on the maps to other sources. The red circles in Figure 15.4b show two regions with exceptionally frequent hop-growing farms and my study is concentrated on these two areas.

II.4. Supplementary sources

In order to evaluate the maps' value as a source, I have compared them to data from contemporary tax registers. I assume that a high degree of correlation between parishes paying a large part of their tax in hops and those showing a high frequency of hop gardens on the maps will strengthen the source value of the map. The same applies, of course between low or negligible amounts of hops in the tax registers and infrequent occurrence of hop-gardens on the maps. Not all tax registers mention hops, and the tax that was actually paid could, depending on different tax reductions, be lower than the amount mentioned in the registers, but I expect to be able to see if the two sources correspond.

I have selected tax registers as close as possible to the maps in time, covering parishes in 'hop-intensive' as well as 'hop-less' regions according to the maps. The selected parishes must include a sufficient number of farms to be useful for this purpose. The question is whether the frequent, as well as the less frequent, hop-producing parishes, according to the maps, correspond with information in the tax registers. I have chosen six parishes in Uppland that fulfil the criteria mentioned above. Useful tax registers exist from 1599-1627, which is only 10 to 40 years before the surveying occurred. (National Archives. CD 2740-2741, 1599; CD 2809, 1627). Tolfta and Tierp parishes were surveyed in 1640-1641 by Sven Månsson (map collection A3), and were situated within Tierps hundred, one of the two distinct hop-farming concentrations. A large share of the mapped hamlets/single farmsteads had at least one hop-garden. Edebo and Knutby parishes were surveyed 1640-1641 by Sven Månsson (map collection A3). The parishes were situated outside the hop-growing concentrations, and the surveyor noted only a few hop-gardens. Hagby and Balingsta parishes (Hagunda hundred), lack hop-gardens according to Sven Månsson's and Johan Botvidsson's maps from 1635 and 1636 (map collection A4).

The tax registers show that the amount of hops paid for was about 4 markpund (about 0.4 kg.) per farm in parishes like Edebo and Knutby, where the geometrical maps contain rather few hop-gardens. (Jansson, 1950: 48 gives the metric equivalent for the markpund). The amount is almost twice as great — about 7 markpund — for farms within the hop-producing concentrations on the maps (Tolfta and Tierp). In parishes without mapped hop production (Hagby and Balingsta) the tax registers do not mention hops among the taxable goods.

The tax registers do not tell us where the hops were grown, so we cannot be certain that it was the farm's own hops that that were counted. What we can see, is that where there are few hop-gardens on the maps, the tax registers give only a conventional estimate mention, usually 4 markpund per farm. In the contrasting situation, where the hop-gardens are common on the maps, the registers contain a careful calculation per farm in the parishes and there the amount paid per farm was 7 markpund on average. One interpretation is that in the former case, the amount to be paid per hop plant was estimated rather than being calculated from the farm's actual hop-garden. In the latter case, the amount was probably calculated on the basis of actual hop production.

As shown above, the two sources (maps and tax registers) fit well together, which strengthens the impression that the surveyors registered the actual presence of hop-gardens. The maps can consequently be used to identify regions with frequent, as well as no, hop-gardens. The concentrations seem to be substantive and, consequently, the maps can be considered useful and reliable for this study.

III. Hop-growing: specialization or subsistence farming?

I define hop-specialization as a farm with 500 poles of hops or more. At this level, only a minor portion of the outcome would have been meant for household needs, which ran from 60 to 100 poles, according to the calculations above, so the main purpose was consequently cash-crop production. The first observation is that the two concentrations of hop-producing farms studied (within Västergötland and Uppland, Figure 15.4, above) contained both kinds of farms, subsistence and cash-crop. The subsistence farm was characterized by sufficient areas of arable land and hay-meadows, supplemented by a hop-garden containing between 50 to 150 poles. This kind of farm occurred in both provinces, whilst the specialized farms can be seen only in Västergötland. The Lillavreta farm in Uppland provides an example of a subsistence farm (Figure 15.5). The hop-garden is not shown on the map, but land surveyor Thomas Christiernsson described it as containing one hundred poles. Concentrations of hop-producers in Uppland often look like this, with growths consisting of 100 to 200 poles. The production must have been enough to pay the tax and provide for household needs, but the main reason for growing the crop was obviously not commercial. The average hop-garden in the concentrated hop-growing parish Alunda in Uppland contained only 95 poles.

Figure 15.5. The subsistence farm

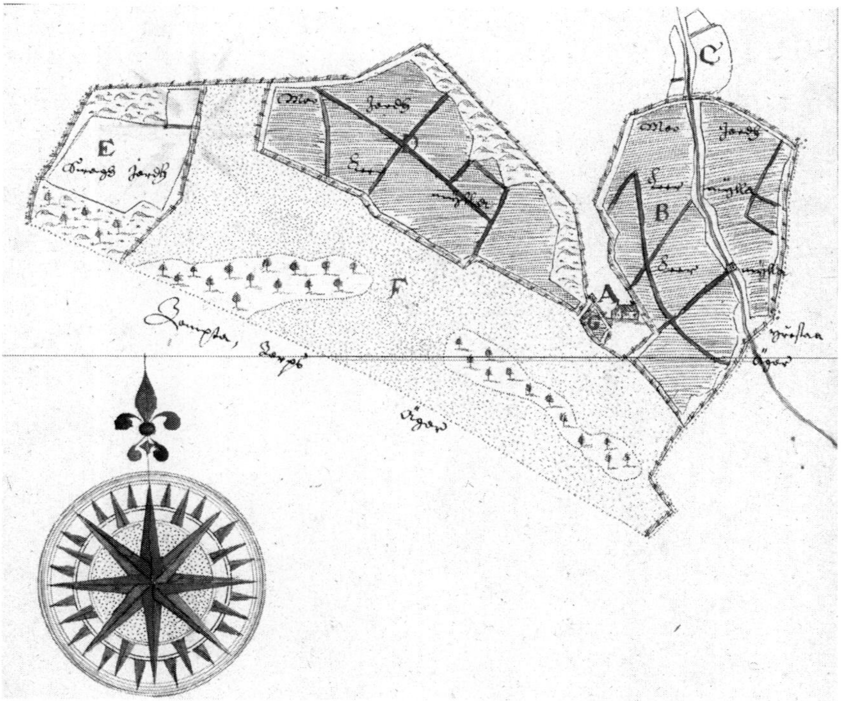

Source. A5: 184, file 092, Lillavreta, Rasbo parish and hundred, Uppland. Surveyed by Thomas Christiernsson, 1640-1641.

Note. The Lillavreta holding illustrates an example of the common hop-growing farm in central Uppland. The farm had a yearly sown area of fifteen acres, a water mill and a hop-garden with one hundred poles. The hop-garden is not shown with a symbol, but mentioned in the text.

The other kind of distinguishable hop-producers can, without hesitation, be defined as specialists, with 500 to 8,000 hop-poles. This amount is a sign of cash-crop production. Several hop-producers in Västergötland had more than 500 poles per farm. Thirty-six of the 88 hop-gardens studied in Västergötland contained 500 poles or more, while 16 hop-gardens had 1,000 poles or more. The Huljesten farm is a good example of specialization, even though, with its 8,000 poles, this was absolutely the largest hop-garden I have come across in the maps (Figure 15.6). This farm could either support nearly 80 families with their hops, or, at a minimum, sell its produce of 300 kilos of dried hops (calculation from Karlsson Strese, 2008b: 47).

Hops are a resource-demanding crop that need a lot of manure and take a great deal of time to tend, to harvest, and to dry, and therefore the economic return had to be satisfactory if the effort was to be profitable. In these matters, hops can be compared to other resource-demanding and profitable crops like vegetables, wine and tobacco. Paul Warde and Niels Grune both reach the same conclusion concerning these crops in Germany before the Agrarian Revolution; they consider them to have been very important sources of cash income. Even the very small landholder, or the landless, could grow hops, wine or tobacco and make a good profit. These cash-crops required greater human effort than grain, but yielded a considerably higher profit (Warde, 2006: 294; Grune, 2007: 71-93).

Figure 15.6. The specialist farm

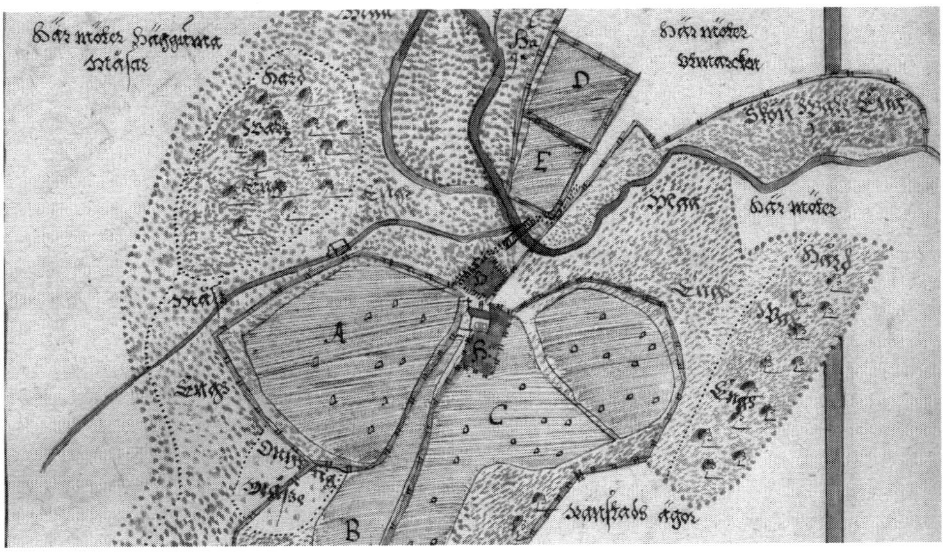

Source. Huljesten, P2: 96 Stenstorps parish, Gudhems hundred, Västergötland. Surveyed by Johan Botvidsson, 1644-1647.

Note. The Huljesten farmstead with its two impressive hop-gardens contained 8,000 poles in all. The gardens are shown as two green, enclosed areas close to the house symbol. The farm used ten acres yearly for grain growing in a three- field rotation system.

There are other distinguishable differences between the hop-growers in the provinces of Uppland and Västergötland. The Uppland farms were more uniform in the size of their hop-gardens, and there is a high frequency of producers within the 'hop-clusters'. The average hop-garden consisted of between 100 and 200 hop-poles, and hops were often mentioned in the tax registers. The Västergötland farms show greater diversity. Hamlets in this region often consisted of farms with no hop-gardens side by side with farms with large ones, containing some 1,000 poles. Individual farms could grow between 800 and 4,000 poles (8,000 in the case mentioned above), and the average hop-garden had 500-750 poles. Hops occurred very rarely in Västergötland tax registers.

There is nonetheless an obvious similarity between the two regions, and in their degree of specialization. Among the total of 1,100 farmsteads I have studied in different parts of rural Sweden, I have never seen a hop-specialist totally dependent on the crop. By dependent I mean landless hop growers, or farmers who reduced their grain-producing area in order to concentrate on this single crop. On the contrary, it seems as if the households' needs were always supplied first and foremost. Paul Warde came to the same conclusion (Warde, 2006: 294) in his discovery that even the large farms in seventeenth-century south-western Germany always met their own needs first, before considering selling their production.

IV. Discussion

The main question of this paper, whether the presence of hop-gardens has anything to do with agricultural specialization, has been answered in the affirmative. The studied maps and tax registers have shown that there was, regionally, a well-developed cash-crop specialization in Swedish agricultural society in the early seventeenth century. The minimum size of a hop-garden to be considered as specialization has been put at 500 hop poles. The household needs, including tax payment in hops, has been calculated at 100 hop poles. There are two distinguishable categories of hop-growing farms, according to the geometrical maps. On the one hand, there are the farms with a small hop-garden of fifty to one hundred poles, more or less enough for household needs. On the other hand, we find the kind of farm mainly discussed here; farms with sufficient arable and hay-meadows as well as large-scale hop- gardens with 500 to 8,000 hop poles. The production of hops among these farms well exceeded household needs, and I interpret this as economic specialization based on hops as a cash-crop.

I see, however, no indication of farms exhibiting signs of the 'modern' form of specialization that involves producing only one crop, and using the profits to buy the farm's daily necessities. What the maps do show is an early form of specialization, characterized by production of hops as a cash crop in excess of the ordinary production of grain and hay together, combined with animal husbandry in a framework of subsistence farming. Even when the hop-gardens were very large and hence time-consuming, they did not replace the acreage needed for subsistence production, or affect the area the latter required.

The explanation of why some regions developed a specialization in hops could be that there was a large regional demand from growing towns, and a strong demand for beer to supply the large gatherings of armed men mustered during this time of wars. However, this is not the only answer; several other regions with similar geographical conditions and in a similar proximity to a town did not display the same development. A long, local tradition of hop-growing must have played a very important role in laying the foundation for the well-developed transportation and purchasing networks that made extensive and work-intensive hop farming profitable in the early seventeenth century.

Hop-growing traditions appear to have followed different paths. In central Uppland the hop-farming tradition was old, medieval at least, hop-growing farms were frequent and the growing areas middling or small in size, ranging from 50 to 200 or 300 poles. No specialist farms were found here. There was a long tradition of hops as a part of taxation in Uppland. In Västergötland there was a different tradition, maybe newer, possibly a sign of the developing individualized and market-oriented farming. There was a substantial difference between subsistence hop growers and specialists in this region, specialized hop growers were frequently met with. There were often large differences within the same hamlet, with some neighbouring farms growing no hops at all and others with a couple of thousand poles' worth of them.

Archives

National Archives *Riksarkivet* Stockholm, Upplands handlingar. CD 2740-2741, år 1599; CD 2809 år 1627.

Database

Project National edition of the oldest geometrical maps. Information about the project and the database (in progress) can be obtained on www.statensarkiv.se/geometriska.

Maps

Project National edition of the oldest geometrical maps.

Scanned versions of geometrical map collections P2 and Öresten, Västergötland. A1; A2; A3; A4; A5; A8; A10 and Smedbygodsets map collection, Uppland. Information about the project and the database (in progress) can be obtained on www.statensarkiv.se/geometriska.

Bibliography

1734 års landslag. Ur: Sveriges Rikes lag (The Swedish Law 1734). Gillad och antagen på Riksdagen år 1734. Serie 1, Rättshistoriskt bibliotek, 37 (Published 1984), 398 p.

BAIGENT, Elizabeth (2005), 'Monarchs, Ministers, and Maps: The Geometrical Mapping of Early Modern Sweden in International Perspective', *in* Birgitta ROECK Hansen (ed.), *Nationalutgåva av de äldre geometriska kartorna,* Stockholm, Kungl. Vitterhets historie och antikvitetsakademien. Konferenser 57, Motala, p. 23-60.

DOVRING, Folke (1953), *Agrarhistorien*, Stockholm, Hugo Gebers, 127 p.

EKSTRAND, Viktor (1901), *Samlingar i Landtmäteri. Första samlingen Instruktioner och Bref 1628-1699,* Stockholm, Isaac Marcus' Boktr.-Aktiebolag, 264 p.

GRÜNE, Niels (2007), 'Commerce and Community in the Countryside: The Social Ambiguity of Market-Oriented Farming in pre-Industrial Northern South-West Germany c. 1770-1860', *Rural History,* 18, p. 71-93.

HELMFRID, Staffan (1959), 'De geometriska jordeböckerna – 'skattläggningskartor? (Was taxation the reason behind the Swedish maps ?), *Ymer*, 3, p. 224-231.

JANSSON, Sam Owen (1950), *Måttordbok; svenska måttstermer före metersystemet*, Stockholm, Nordiska Museet, 109 p.

KARLSSON STRESE, Else-Marie (2008a), 'Humle finns i öl av 'fel' skäl' (Hops in Beer by 'Wrong' Reasons), *Forskning och Framsteg,* 3, p. 30-35.

KARLSSON STRESE, Else-Marie (2008b), 'Humlegårdar på kartor, växter i naturen', *in* Mats HÖGLUND (ed.), *1600-talets jordbrukslandskap,* Stockholm, Riksarkivet, 9 p.

TOLLIN, Clas (1991), *Ättebackar och ödegärden: de äldre lantmäterikartorna i kulturmiljövården,* Svenska, För vuxna, 96 p.

TOLLIN, Clas (2003), 'Den svenska storskaliga karteringen 1633-1655', in Birgitta ROECK HANSEN (ed.), *Nationalutgåva av de äldre geometriska kartorna,* Kungl. Vitterhets historie och antikvitetsakademien, Konferenser 57, Motala 2005, p. 6-22.

TOLLIN, Clas (2007), 'När Sverige sattes på kartan' (When Sweden was Put on the Map), in Ulf JANSSON (ed.), *Kartlagt land*, Stockholm, Kungl. Skog- och Lantbruksakademiens

WARDE, Paul (2006), 'Subsistence and Sales', *Economic History Review*, 59, p. 289-319.